# THEORETICAL MORPHOLOGY

*Perspectives in Paleobiology and Earth History*

*Critical Moments in Paleobiology and Earth History Series*
David J. Bottjer and Richard K. Bambach, Editors

J. David Archibald, *Dinosaur Extinction and the End of an Era: What the Fossils Say*

George R. McGhee, Jr., *The Late Devonian Mass Extinction: The Frasnian/Famennian Crisis*

Betsey Dexter Dyer and Robert Alan Obar, *Tracing the History of Eukaryotic Cells: The Enigmatic Smile*

Donald R. Prothero, *The Eocene-Oligocene Transition: Paradise Lost*

Douglas H. Erwin, *The Great Paleozoic Crisis: Life and Death in the Permian*

Mark A. S. McMenamin and Dianna L. S. McMenamin, *The Emergence of Animals: The Cambrian Breakthrough*

*The Perspectives in Paleobiology and Earth History Series*
David J. Bottjer and Richard K. Bambach, Editors

Anthony Hallam, *Phanerozoic Sea-Level Changes*

Ronald E. Martin, *One Long Experiment: Scale and Process in Earth History*

Judith Totman Parrish, *Interpreting Pre-Quaternary Climate from the Geologic Record*

The concept for these series was suggested by Mark and Dianna McMenamin, whose book *The Emergence of Animals* was the first to be published.

# Theoretical Morphology

## THE CONCEPT AND ITS APPLICATIONS

GEORGE R. McGHEE, Jr.

COLUMBIA UNIVERSITY PRESS

NEW YORK

Columbia University Press
Publishers Since 1893
New York   Chichester, West Sussex
Copyright © 1999 Columbia University Press
All rights reserved


Library of Congress Cataloging-in-Publication Data

McGhee, George R.
  Theoretical morphology : the concept and its applications / George R.
  McGhee, Jr.
      p.   cm. — (Perspectives in paleobiology and earth history)
    Includes bibliographical references and index.
    ISBN 978-0-231-10616-0 (alk. paper). — ISBN 978-0-231-10617-7 (pbk. : alk. paper)
      1. Morphology—Mathematics.   2. Morphology—Statistical methods.
  3. Morphogenesis—Mathematical models.   I. Title.   II. Series: Perspectives in
  paleobiology and earth history series.
  QH351.M335   1998
  571.3'01'51—dc21                                                      98-8660

Casebound editions of Columbia University Press books are printed on permanent
and durable acid-free paper.
Printed in the United States of America

To MARAE,
*A Mharae, mo ghràidh.*

# CONTENTS

This book is an attempt to bring to the reader some of the fascinating developments that have occurred in the past three decades in the field of theoretical morphology and to point the way to research areas of great promise that have yet to be explored. The future of theoretical morphology lies with the graduate students of the present and future generations. With that fact in mind, this book is specifically written for the graduate student considering a career in morphology or the professional morphologist with graduate students hunting for research topics. So if you are a graduate student considering research in the field of morphology, please take note, as challenging potential Ph.D. dissertation topics abound in the chapters to come.

The modern field of theoretical morphology was founded by Professor David M. Raup in a series of classic papers between 1961 and 1968, published in *Science*, the *Journal of Paleontology*, and the *Proceedings of the National Academy of Sciences*. Those classic papers are just that, classic, and have continued to be cited over and over in the past 30 years. In recent years, however, some paleobiologists and biologists have come to believe that those classic papers have been, in fact, cited too much. For example, the modern morphologist Savazzi (1995:238) maintains that most university studies "include, at most, a brief mention of Raup's work, typically accompanied by a few slides taken from Raup's original illustrations," as if the field of theoretical morphology has not changed since the 1960s. Many very interesting developments have occurred in the field since those early days of Raup's original work, and the promise of theoretical morphology to solve some of our most perplexing problems in evolutionary paleobiology has yet to be fully realized. The exploration of these interesting new developments, as well as the examination of the perplexing problems that still face us, is the subject of this book.

Dave Raup was my Ph.D. mentor, and my doctoral dissertation itself concerned the theoretical morphology of biconvex brachiopods. In 1991 I was asked by the Paleontological Society to write an overview of the field of theoretical morphology and to present a paper in the society's *Short Course in Analytical Paleobiology* at the annual meeting of the Geological Society of America. That paper was very well received, and at the meeting Dave Raup strongly urged me to consider writing a more extensive book on the subject. That request has had to wait.

The two great loves of my scientific life have been morphology and ecology. That is, the study of the many forms that animals and plants have produced in their long evolution on this planet and the interrelationships of those same animals and plants with one another and their environment through geologic time. In 1996 I concluded a long and fascinating journey exploring the ecology of the Late Devonian mass extinction (see McGhee 1994 for a very short summary and McGhee 1996 for a much more extensive treatment of this biotic crisis, one of the "Big Five" in Earth history). Thus in 1997, some six years later, I now have the time to turn my attention back to morphology and to follow Dave's request and recommendation to write this book on the field of theoretical morphology.

I would like to acknowledge, with gratitude, my graduate training with David M. Raup and Daniel C. Fisher, two very fine and also very different morphologists. I would also like to acknowledge the many discussions and morphologic musings I have had with Ulf Bayer, Adolf Seilacher, Wolf-Ernst Reif, and Jürgen Kullmann during the years in which I worked with the University of Tübingen's Konstruktions morphologie research group. I also thank all those my colleagues who have argued with me about brachiopod morphology for many years, who are so numerous I cannot begin to name them all.

And finally, I thank my wife, Marae, for her patient love.

# THEORETICAL MORPHOLOGY

PHYSIOLOGICAL MORPHOLOGY

# What Is Theoretical Morphology?

Depending on your point of view, theoretical morphology is either a very new field of study or a very ancient one. The term "theoretical" morphology was first used by E. S. Russell in his classic book *Form and Function: A Contribution to the History of Animal Morphology* more than eight decades ago (Russell [1916] 1982:33). In addition, Lauder (1982:xviii) argued:

> *Form and Function* demonstrates the antiquity of many issues being debated today. For example, the notion that extant organismal forms are only a subset of the range of theoretically possible morphologies (referred to as theoretical morphology; see Raup and Michelson 1965, McGhee 1980[b]) can be found in the writings of Cuvier.

It is also true that theoretical geometric models of natural morphology are not new, and neither is the debate over whether model or parameter is biologically more appropriate (e.g., the debate between Moseley [1838, 1842] and Naumann [1840a,b; 1845]).

What is understood as theoretical morphology today, however, was generally not feasible until the advent of the computer. Two quite different conceptual areas of evolutionary biology are understood today under the umbrella term of theoretical morphology: (1) the mathematical simulation of organic morphogenesis and (2) the analysis of the possible spectrum of organic form via hypothetical morphospace construction. The second concept follows from the first but has quite different goals.

In the first conceptual area, the initial focus in the early days was to model existent form with a minimum number of parameters and mathematical complexity. In essence, the goal was to reveal general and common geometric principles in what, at first glance, might appear to be widely divergent organic morphologies (such as those exhibited by ammonites and brachiopods). More recent work in this area of theoretical morphology models the actual process of biological morphogenesis itself.

In the second conceptual area, the goal is to explore the possible range of morphologic variability that nature could produce by constructing $n$-dimensional geometric hyperspaces (termed "theoretical morphospaces"), which can be produced by systematically varying the parameter values of a geometric model of form. Such a morphospace is produced without any measurement reference to real or existent organic form. Once constructed, the range of existent variability in form may be examined in this hypothetical morphospace, both to quantify the range of existent form and to reveal nonexistent organic form. That is, to reveal morphologies that theoretically could exist (and can be produced by computer) but that never have been produced in the process of organic evolution on the planet Earth. The ultimate goal of this area of research is to understand why existent form actually exists and why nonexistent form does not.

The founder of modern theoretical morphology was David M. Raup, who first used the term in 1965 in a paper written with A. Michelson. Although Raup was using theoretical morphology in the first conceptual sense in his earlier papers (Raup 1961, 1962), the term theoretical morphology itself was originally used in only the second conceptual sense, as can be seen in the brief (two-sentence) abstract of their classic *Science* paper "Theoretical Morphology of the Coiled Shell": "In studying the functional significance of the coiled shell, it is important to be able to analyze the types that do not occur in nature as well as those represented by actual species. Both digital and analog computers are useful in constructing accurate pictures of the types that do not occur" (Raup and Michelson 1965:1294). Later, however, Raup explicitly expanded the usage of the term theoretical morphology to cover modeling or simulation of the actual process of biological morphogenesis itself, as seen in his paper "Theoretical Morphology of Echinoid Growth": "The computer model also has little biological basis, but, because it succeeds in simulating morphology (and ontogenetic development), we may conclude that the *actual biological system controlling plate patterns need not be more complicated than that used in the computer simulation*" (Raup 1968:62; italics Raup's).

## THE DISCIPLINE OF THEORETICAL MORPHOLOGY AND ITS GOALS

*Morphospace is potentially one of paleontology's most significant contributions to the analysis of form. It is the tool by which we can document the range of actual structures that have evolved in the history of life as a subset of the structures that are theoretically possible.*

*Hickman (1993b:170)*

Of the two research goals of theoretical morphology—the creation of theoretical models of morphogenesis and theoretical morphospace analyses of the evolution of organic form—it is the latter that has received more attention and that has been hailed as the greater contribution to our understanding of how evolution works. I have therefore decided to consider theoretical morphospaces first. In chapter 2 we shall examine in detail the fundamental concept of the theoretical morphospace and how a theoretical morphospace is constructed. Then, in chapters 3 through 7, we shall examine actual case studies of theoretical morphospace analyses. Only in chapter 8 will we consider the goal of theoretical models of morphogenesis, first with an example in that chapter of how mathematical models are constructed and then with the examination of the many different theoretical morphogenetic models in chapters 9 and 10.

We shall now approach the question, What is theoretical morphology? from a different perspective by considering what it is *not*. Theoretical morphology is not morphometrics or functional morphology or constructional morphology.

## THEORETICAL MORPHOLOGY AND MORPHOMETRICS

*Theoretical morphology differs from biometrics or biostatistics as it searches for the essential geometric growth rules in any particular case.*

*De Renzi (1995:241)*

Morphometrics is a school of morphologic analysis concerned with the precise measurement of form in individual organisms and with

3

the precise comparison of those measurements among different individuals. It is the school of morphometrics that is most frequently confused with theoretical morphology. This confusion is likely due to the fact that both morphometrics and theoretical morphology are highly mathematical and that both schools extensively use computers. However, the goals and techniques of theoretical morphologic analyses and morphometric analyses are profoundly and fundamentally different.

Because morphometrics is concerned with the precise quantification of existent morphology, it often seeks to reproduce form as a sort of mathematical picture by means of various mathematical techniques of shape analysis and outline-fitting algorithms. But morphometrics does not simply seek to precisely portray existent morphology via quantification; it is even more concerned with the precise quantitative comparison of multiple existent morphologies which may vary in form to a greater or lesser degree. Thus morphometric analyses are often characterized by the intensive examination of existent morphology to discern homologous features from analogous ones and to separate synapomorphic characters from symplesiomorphic ones. And that is extremely important if we are to obtain an accurate picture of the morphologic changes that have taken place in the evolution of a group of organisms and if we are to construct phylogenies for those organisms that are holophyletic, phylogenies that accurately reflect ancestor-descendant relationships.

An early inspiration for morphometrics was D'Arcy Thompson's coordinate transformation method of shape comparisons (1917, 1942). Over the many years since his work, morphometricians have devised even more precise methods of shape quantification and comparison, from simple bivariate plots to multivariate hyperspaces, from Fourier series to fractals (Mandelbrot 1983; see Lutz and Boyajian 1995). The morphometric literature is enormous and cannot be summarized here. Suffice it to say simply that morphometrics is fundamentally concerned with quantification.

In contrast, theoretical morphology is interested in simulation, not in quantification. Theoretical morphology is concerned with the simulation of the principal aspects of form with a minimum number of geometric parameters, or with the simulation of the morphogenetic process itself that produced the form under study, and is not concerned with the production of a precise mathematical characterization or picture of any given existent form. In fact, the creation and examination of *nonexistent form* is often of more interest in theoretical morphologic analyses than the examination of existent form.

The second major goal of theoretical morphology is the construction of $n$-dimensional geometric hyperspaces called theoretical morphospaces. The mathematical techniques of morphometrics have also been used to make hyperdimensional morphospaces, typically by applying ordination techniques to matrices of measurements taken from existent organisms. These hyperspaces I have termed "empirical morphospaces" (McGhee 1991), as opposed to "theoretical morphospaces," as they are fundamentally different from theoretical morphospaces. To mention just one such difference, an empirical morphospace does not exist in the absence of measurement data, whereas a theoretical morphospace does. The differences between these two types of morphospace will be explored in greater detail in chapter 2.

## THEORETICAL MORPHOLOGY AND
## FUNCTIONAL MORPHOLOGY

*Although theoretical morphology does not play a direct role in the inference of function in fossils, it can be used to generate and organize functional hypotheses and to compare the range of designs that have evolved with what is theoretically possible.*

*Hickman (1988:790)*

Functional morphology is a school of morphologic analysis that specifies the functional or adaptive aspects of organic form, usually by employing the paradigm methodology of Martin J. S. Rudwick (1964; see also the review in Lugar 1990). The paradigm methodology is a formalized technique of mechanical analogy in investigating the functional significance of an organic structure or form. Several functions are first postulated for the structure under investigation, and then the optimal mechanical design for each function is determined (the paradigm). The paradigm form that most closely matches the original structure is considered to demonstrate the actual function of the structure.

It cannot be denied that much of organic form has adaptive significance and represents an adaptation. That fact does not mean that all organic form is adaptive, however, as we shall see in the next section, on constructional morphology. Using the paradigm methodology to analyze the adaptive significance of organic form has been sharply criticized by some (Gould and Lewontin 1979), and the very logic of the methodology itself has been declared invalid by some (Signor 1982) and carefully validated and justified by others (Fisher 1985).

Functional morphology is used in the latter stages of a theoretical morphospace analysis when the adaptive significance, if any, of the distribution of real organic form in a theoretical morphospace of hypothetical form is examined (to be discussed in chapter 2). But functional morphology is generally not used at all in the theoretical modeling of morphogenesis. That is, functional morphology is a separate and valuable tool that may or may not be used in theoretical morphologic analyses.

## THEORETICAL MORPHOLOGY AND CONSTRUCTIONAL MORPHOLOGY

*Bautechnischer Aspekt: . . . sucht die bautechnische Deutung jeweils nach einem modellhaften morphogenetischen "Programm," in dem die beobachtete Form auf wenige Parameter . . . zurückgeführt wird. Das Zeitalter des Computers legt solches Denken nahe und eröffnet sogar die Möglichkeit, gegebene ebenso wie unrealisierte Formen im Einzelfall zu simulieren. ("Theoretische Morphologie," vgl. Raup and Michelson 1965, Raup and Seilacher 1969)*

*Seilacher 1970:395*

Constructional morphology is a school of morphologic analysis founded by Adolf Seilacher (1970; see also the review in Thomas 1979). In constructional morphologic analyses, organic form is postulated to result from the interplay of three constraints or limiting factors: functional (adaptational) constraints, fabricational (morphogenetic) constraints, and historical (phylogenetic) constraints. At the time (in the late 1960s) the school of constructional morphology was founded, this threefold concept of form was a view of morphology radically different from that held by most American and British morphologists, who tended to see all morphology as the result of adaptation (a point of view that came to be termed "hyperselectionist" by Raup [1972] and "adaptationist" by Gould and Lewontin [1979]; see also the review in Gould [1995]). And indeed, the fundamental assumption of the school of functional morphology is that morphology does have a function.

In contrast, Seilacher (1970, 1973) suggested that many aspects of morphology might be side effects of the process of morphogenesis itself, the "fabricational noise" of growth constraints, and not the result of adaptation. In addition, Seilacher (1970) proposed that other aspects of morphology might be the result of phylogenetic constraint, the result of genetic legacy, and also thus not the result of adaptation. Both morphogenetic constraints and phylogenetic constraints are rooted in the concept that organisms possess an evolutionary *Bauplan. Bauplan* is an engineering German term that has been variously translated as "ground plan," "blueprint," or "engineering design." In biology, the term is used to represent the basic architectural and organizational pattern shared by the members of a monophyletic clade of higher taxonomic rank. To quote Valentine (1986:209): "At the upper levels of the taxonomic hierarchy, phyla- or class-level clades are characterized by their possession of particular assemblages of homologous architectural and structural features; in this paper, it is to such assemblages that the term *Bauplan* is applied."

The definition of the term *Bauplan* is not tied to any specific taxonomic rank, and so its usage varies from clade to clade and to subsets of clades within clades. Valentine (1986), for example, uses the term *Bauplan* to refer to the assemblage of architectural features shared by organisms at the phylum or class level, and within these *Baupläne* he uses the term *Unterbaupläne* to refer to distinctive major subtaxa at the class or ordinal level, the next taxonomic level down from the initial designation of the *Bauplan*.

It is generally agreed that *Bauplan* refers to an assemblage, or complex, of architectural features and not to just one distinctive morphologic feature characteristic of a particular higher taxon. Because I am well aware of that general agreement, I must, at this point, take exception to Hall, who wrote: (1996:226)

> For others, the term *Bauplan* is applied to a single morphological structure within a group, as in the morphology and growth of brachiopod valves (McGhee 1980b) or the circuitry of the nervous system (Ebbesson 1984). However, this is not consistent with the essence of the concept of a suite of characters that unites members at higher taxonomic levels. Woodger (1945) discussed this issue and clearly came down against *Baupläne* applying to individual elements.

In reference to this study of the morphology and growth of brachiopod valves (which is covered in chapter 5), I specifically stated: "The planispiral shell of the Brachiopoda can be viewed as an integral part of the *Bauplan* of the phylum, its basic engineering design" (McGhee 1980b:62). That is, the shell is *an integral part* of the *Bauplan* of the phylum Brachiopoda, not the *Bauplan* itself. The concept of the *Bauplan* of the phylum Brachiopoda as an assemblage of features is made even more explicit in McGhee (1980b:62–63):

> Some aspects of the articulate brachiopod *Bauplan* which will be of importance in subsequent discussions are: (1) shell growth by terminal accretion without major modification after deposition of shell material, (2) the presence of two valves in articula-

7

tion, (3) the lophophorate filtering system, with its unfused filaments and mantle edges, and (4) a sessile mode of life

that considers, respectively, mode of growth, the skeleton, mode of feeding and soft tissue characteristics, and mode of life as aspects of the brachiopod *Bauplan*. If anything, I might be criticized for including the last aspect in the *Bauplan*, as it is more ecological than morphological in nature and thus makes the concept of *Bauplan* too broad.

In a constructional morphologic analysis, Seilacher (1970) clearly identified the school of theoretical morphology with the analysis of the morphogenetic, or fabricational, constraints on organic form (*Bautechnischer Aspekt*). And as can be seen in the quotation from his classic paper at the beginning of this section of the chapter, he clearly indicated that both aspects of theoretical morphology (morphogenetic simulation and morphospace construction) were to be so considered. Likewise, Seilacher (1970) clearly identified the school of functional morphology with the analysis of the functional, or adaptational, constraints on organic form in a constructional morphologic analysis. The school of constructional morphology was thus envisioned as a synthetic one, incorporating the disciplines of functional and theoretical morphology and the German concept of evolutionary *Bauplan*.

Since Seilacher's (1970) early visualization of form as a result of the interplay of these three factors, other morphologists have added additional factors (Raup 1972, Hickman 1980). Seilacher (1991) himself expanded the concept of constructional morphology to a concept of "biomorphodynamics" by adding an environmental factor to his previous three morphologic ones.

As a graduate student in 1977, after my first stay of research at the University of Tübingen with the *Konstruktionsmorphologie* research group and many conversations with Seilacher, I wrote in my dissertation: "The synthetic school of constructional morphologic analysis encompasses the paradigm methodology of Rudwick (1964, 1968) and the theoretical morphologic approach of Raup and Michelson (1965)." Dave Raup, my doctoral mentor, read this assessment and made no objection. After more than two decades of thought and research, however, I now view constructional morphology as more a heuristic concept, a working hypothesis, than a specific analytic methodology. In my experience, most morphologists concentrate on the three separate aspects of constructional morphology separately rather than synthesizing them together. That is, functional morphology has remained functional morphology, and theoretical morphology has remained theoretical morphology.

## THE CURRENT STATE OF THE SCIENCE
## OF MORPHOLOGY

*The ultimate triumph of theoretical morphology would be an understanding of biological diversity, framed in terms of the boundaries between the possible and the actual and the possible and the impossible. It should integrate across all levels of structure, from organic molecules to entire and seemingly complex functioning organisms, where as yet undiscovered laws of structural consonance may exist.*

*Hickman (1993b:170)*

The ultimate triumph of theoretical morphology is not even in sight! Consider this optimistic prediction, made almost three decades ago: "A science of form is now being forged within evolutionary theory. It studies adaptation by quantitative methods . . . it seeks to reduce complex form to fewer generating factors and causal influences" (Gould 1970:77). Gould was not referring to morphometrics when he mentioned "quantitative methods" but, rather, the techniques of theoretical morphology and their power in the study of adaptation (or nonadaptation, for that matter). And even though considerable progress has been made in theoretical morphology in the past three decades, it is a fact that more quantitative morphologists today pursue morphometrics than theoretical morphology.

Morphometrics is a very valuable and important school of morphologic analysis, with its own specific goals and techniques. In contrast with popular field of morphometrics, the analytic techniques of theoretical morphology have been sadly neglected. One of the goals of this book is to make clear to the reader the conceptual difference between the goals and techniques of morphometrics and those of theoretical morphology and to reveal to the reader those aspects of the evolution of life on Earth that are ideally suited to theoretical morphologic analyses, but not to morphometric analyses.

In the last chapter of this book, I shall outline a challenge, particularly to graduate students, for the development of theoretical morphology in the future. There are so many interesting things to be done in the field, and we have barely begun to understand the evolution of form on Earth. Many excellent Ph.D. dissertations and intellectually stimulating life careers simply await the attention of bright young graduate students.

# II

# The Concept of the Theoretical Morphospace

*What, then, do we need? Not simply a good method for the multivariate description of organisms, for such we have (Bookstein 1977a,b; Sneath and Sokal 1973). And not even a proper multivariate description of morphological transformation—whether by D'Arcy Thompsonian (1917) coordinate transformation, Huxleyan (1932) allometric growth gradients, or more modern methods like trend surface analysis (Sneath 1967). We need, instead, to define a full range of the abstract (and richly multivariate) space into which all organisms may fit (the morphospace). We must then be able to characterize individual organisms and plot them within this encompassing space.*
*—Gould (1991:420)*

In the previous chapter I mentioned that of the two research goals of theoretical morphology—the creation of theoretical models of morphogenesis and theoretical morphospace analyses of the evolution of organic form—it is the latter that has received more attention and that has been hailed as the greater conceptual contribution to our understanding of how evolution works. The modern concept of the theoretical morphospace is grounded in, and is an extension of, an older hyperspace concept variously termed a "fitness landscape" or an "adaptive landscape." I shall begin this chapter by looking at a concept that predates the modern discipline of theoretical morphology but nevertheless is a concept that was an important catalyst in creating the idea of a theoretical morphospace.

## THE CONCEPT OF THE ADAPTIVE LANDSCAPE
*The idea of a fitness landscape was introduced by Sewall Wright (1932), and it has become a standard imagination prosthesis for evolutionary theorists. It has proven its value in literally thousands of applications, including many outside evolutionary theory.*
*Dennett (1996:190)*

The rationale for constructing an $n$-dimensional hyperspace of theoretical morphologic forms is rooted in Sewall Wright's fitness or adaptive landscape concept (1932), although in theoretical morphology, not all existent morphology is necessarily viewed as adaptive. Wright envisioned a hyperspace of all possible combinations of genes present in different organisms, a multidimensional space that contained not only the genetic compositions of all existing organisms but also the genetic compositions of organisms that do not exist. Why are some genetic compositions represented by living organisms, and others not? To answer this question, Wright postulated a selectionist explanation for the presence or absence of actual organisms with differing genetic compositions, a nonrandom distribution of differential fitness within the genetic hyperspace. Thus was born Wright's concept of a fitness or adaptive landscape, as perhaps best described by his fellow geneticist Dobzhansky (1970:25–27):

> Suppose there are only 1000 kinds of genes in the world, each gene existing in 10 different variants or alleles. Both figures are patent underestimates. Even so, the number of gametes with different combinations of genes potentially possible with these alleles would be $10^{1000}$. This is fantastic, since the number of subatomic particles in the universe is estimated as a mere $10^{78}$. . . . Some gene combinations, indeed a vast majority of them, would be discordant and inviable in any environment. Other combinations, perhaps a tiny minority of the potentially possible ones, are suitable for life in some environments.
>
> This situation can be envisioned with the aid of a symbolic picture of adaptive "peaks" and "valleys" first contrived by Wright (1932). With two gene loci each having 10 alleles, 100 combinations are possible. They can be diagrammed as a two-dimensional grid, on which each combination will be represented by a point. Now, some of the combinations will have a higher and others a lower Darwinian fitness or adaptive value . . . fitness may be depicted as a third dimension, giving the diagram the appearance of a topographic map.
>
> The diversity of living forms may then be envisaged as a multitude of adaptive peaks, corresponding to the multitude of ways of living that are possible on our planet. The variety of these possible ways of living—ecological niches—is, however, not only great; it is also discontinuous. . . . Hence, the living world is not a formless mass of randomly combining genes and traits, but a great array of families of related gene combinations, which are

clustered on a large but finite number of adaptive peaks. Each living species may be thought of as occupying one of the available peaks in the field of gene combinations. The adaptive valleys are deserted and empty.

A slightly different graphic representation of an adaptive landscape is given in figure 2.1. Because the topic of this book is the analysis of morphology—the phenotype rather than the genotype—I have substituted "morphocharacter 1" and "morphocharacter 2" for the horizontal dimensions of the three-dimensional surface given in the figure, rather than use "alleles at gene locus 1" and "alleles at gene locus 2," as Dobzhansky (1970) did. The vertical dimension in figure 2.1 is "degree of adaptation" or "fitness." By following the gridlines across the three-dimensional surface, the reader can discern which combinations of morphocharacters 1 and 2 have the highest or optimum degree of adaptation (the two peaks), which morphologic combinations have less than optimum but perhaps still viable adaptations (the slopes on

FIGURE 2.1. A hypothetical adaptive landscape of morphologic combinations (x-y axes) versus fitness or the degree of adaptation (z-axis). Topographic highs represent adaptive morphologies that function well in natural environments (and therefore are selected for), and topographic lows represent unadaptive morphologies that function poorly in natural environments (and therefore are selected against). *Modified from McGhee 1980b.*

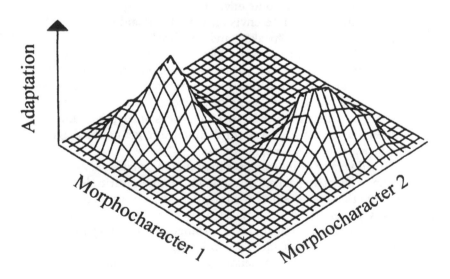

the two peaks), and which morphologic combinations are lethal, with zero adaptive potential (the flat plane). The space between the two peaks in figure 2.1 thus represents an adaptive valley.

Although the adaptive landscape concept has been termed a "standard imagination prosthesis for evolutionary theorists" (Dennett 1996), the many uses of the concept have been mostly just that, imaginary or conceptual. The actual utilization of an adaptive landscape in the actual analysis of genotypes or morphologies is hampered by the complexity of biological form, and for that reason some critics have stated that the adaptive landscape concept is of heuristic value only. In theoretical morphology, however, the concept is put into actual practice, as we shall see.

## DEFINITION OF THE THEORETICAL MORPHOSPACE

*The study of evolutionary constraint requires a metric—a map for visualizing the occupied and unoccupied evolutionary pathways that are theoretically possible. Two such maps are the often cited "adaptive landscape" of genetic frequencies (from Sewall Wright) and David M. Raup's "morphospace" of coiled shells.*

Schindel (1990:270)

At the beginning of the chapter, I stated that the adaptive landscape concept was an important catalyst in the creation of the idea of a theoretical morphospace. We can now ask, What is a theoretical morphospace?" that is, How does the concept of a theoretical morphospace differ from the concept of an adaptive landscape?

Theoretical morphospaces may be defined most explicitly as "*n*-dimensional geometric hyperspaces produced by systematically varying the parameter values of a geometric model of form" (McGhee 1991:87). Adaptive landscapes and theoretical morphospaces share the feature of hyperdimensionality, although the dimensionality of a theoretical morphospace is always much less than that envisioned by Sewall Wright as necessary for an adaptive landscape large enough to encompass all possible combinations of genes present in life on Earth. The crucial concept shared by both theoretical morphospaces and adaptive landscapes is their ability to specify nonexistent form: nonexistent genotypes in the case of the adaptive landscape and nonexistent phenotypes in the case of the theoretical morphospace.

The main difference between the adaptive landscape and the theoretical morphospace lies in their dimensions. The dimensions of the adaptive landscape are gene loci and fitness. If we switch from the genotype to the phenotype, the dimensions of the adaptive landscape

13

are morphocharacters and fitness (as in figure 2.1). The dimension of fitness or degree of adaptation is a fundamental feature of the adaptive landscape concept.

In contrast, the dimensions of a theoretical morphospace are geometric or mathematical abstractions of form. The concept of adaptation does not enter into the construction of a theoretical morphospace, whereas the concept of adaptation is a fundamental feature of an adaptive landscape. Differing positions in an adaptive landscape not only represent differing genetic combinations, they also represent different degrees of adaptation or fitness. Because the dimensions of a theoretical morphospace are mathematical abstractions, differing positions within the theoretical morphospace simply represent morphologies produced by the combination of differing coordinates along the dimensional axes. Given a theoretical morphospace, one can then determine which of the hypothetical forms in that morphospace have actually been produced in nature and which, though theoretically possible, have not. The absence of actual organic forms in a theoretical morphospace does not necessarily mean that the hypothetically possible, but naturally nonexistent, morphologies are nonadaptive (something that would be automatically assumed in an adaptive landscape). An alternative point of view could just as easily maintain that such hypothetical nonexistent morphologies might function perfectly well in nature  but that the process of evolution simply has not produced them yet. This point will be considered in more detail in the next section of this chapter.

Another important feature of the dimensions of a theoretical morphospace is that they are defined without any reference to actual measurement data from existent form. The very ability of a theoretical morphospace to reveal nonexistent form is a function of the measurement-independent nature of the morphospace's dimensions. Theoretical morphospaces of hypothetical form exist in the absence of any morphometric data from real organisms (and are thus very different from empirical morphospaces, which will be shown in a later section of this chapter).

And last, if the dimension of "degree of adaptation" or "fitness" is somehow also mapped into a theoretical morphospace, then the concepts of an adaptive landscape and a theoretical morphospace converge. Actual examples of this convergence will be considered in chapters 3 through 6.

## THE PROBLEM OF CIRCULARITY: MORPHOLOGIC
## PEAKS ARE NOT ADAPTIVE PEAKS

In theoretical morphospace analyses, some care must be taken not to fall into the circular adaptationist trap that "existence in and of itself is the sufficient demonstration of adaptation." It is not. As initially constructed, the theoretical morphospace is not an adaptive landscape.

The analysis of form using theoretical morphospaces is actually a three-phase process: (1) the construction of a theoretical morphospace of hypothetical yet potentially existent morphologies, (2) the examination of the distribution of existent form in the morphospace to determine which forms are common, rare, or nonexistent in nature, and (3) the functional analysis of both existent and nonexistent form to determine whether the distribution of existent form is indeed of adaptive significance.

Phase 1 can be an end in itself. Sometimes it is apparent from simple visual examination of the computer-produced morphologies in a theoretical morphospace where the boundary falls between existent and nonexistent form. Often, however, it is not. Figure 2.2 shows a two-dimensional slice through a four-dimensional theoretical morphospace, which will be explained in detail in chapter 4. The hypothetical morphologies illustrated in figure 2.2 look like real ammonoids but in fact are mathematical constructs. I invite the reader to predict, on the basis of the range of hypothetical form illustrated in figure 2.2, where the form distribution of real ammonoids is. Not easy, is it? Perhaps only a paleontologist trained as a specialist in ammonoid evolution could predict the real range of ammonoid form in the theoretical morphospace on the basis of visual examination of the hypothetical forms alone.

It is therefore usually necessary (and also interesting) to see exactly what nature has produced within the range of potential form and in what frequencies, that is, to proceed to Phase 2. The frequency of individual morphologies in the morphospace is usually portrayed by density contours, which show the density of points (individuals) in the morphospace. The frequency distribution of actual ammonoid morphologies in the theoretical morphospace illustrated in figure 2.2 is given in this contour-map format in figure 2.3. By comparing figures 2.2 and 2.3, the reader can see which potential ammonoid morphologies actually exist in nature and which do not. And of the potential morphologies that do exist, the reader can also see which are most frequently found in ammonids (the peak).

The density contour of the frequency of existent form (figure 2.3) can be deceptively similar to an adaptive peak of fitness contours in Sewall Wright's adaptive landscape concept, introduced at the beginning of this chapter (cf. figures 2.1 and 2.3). However, in a theoretical morphospace analysis, the proof (or disproof) that the most frequently occurring morphology is indeed the most functionally superior mor-

FIGURE 2.2. Computer-produced hypothetical ammonoids in a univalved theoretical morphospace. Although the hypothetical ammonoids look like real ammonoids, they are mathematical constructs and exist in the absence of any measurements taken from actual ammonoids. *From Raup 1967. Used with permission of SEPM—Society for Sedimentary Geology (formerly Society of Economic Paleontology and Mineralogy).*

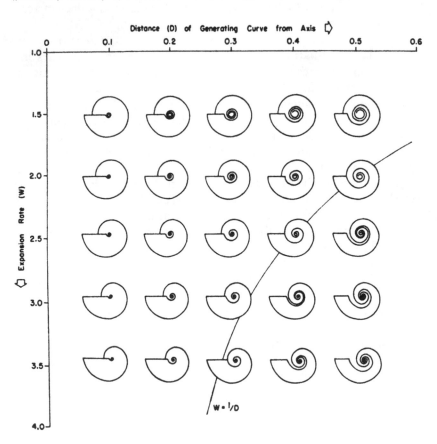

**FIGURE 2.3.** The frequency distribution of morphologies of 405 actual ammonoids in the theoretical morphospace given in figure 2.2. Contours measure the increase in density of genera per unit area of the plot. Note that the "topographic high" in this figure represents the most frequently occurring morphology found in the ammonoids and does not represent the "degree of adaptation" or "fitness" of that morphology, as in figure 2.1. *From Raup 1967. Used with permission of SEPM—Society for Sedimentary Geology (formerly Society of Economic Paleontology and Mineralogy).*

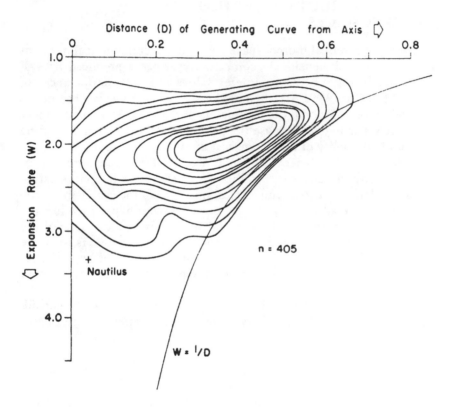

phology is a separate and independent endeavor, Phase 3. As Raup stated (1969), the revelation of the spectrum of existent morphology in the realm of theoretically possible morphology does not in itself interpret the significance of that spectrum, but it does provide a powerful vehicle to facilitate the analysis of that spectrum.

## THE CONSTRUCTION OF THEORETICAL MORPHOSPACES

How is a theoretical morphospace constructed? If theoretical morphospaces are "$n$-dimensional geometric hyperspaces produced by systematically varying the parameter values of a geometric model of form" (McGhee 1991:87), then obviously one first needs a geometric model of form. Second, the parameters of that geometric model must be as few as possible. Because the parameters of the model determine the dimensionality of the theoretical morphospace, the more parameters there are, the higher the hyperdimensionality of the morphospace will be. A hyperspace with 1 million dimensions may be an interesting concept, but it is also an impractical one. One of the criticisms of the concept of the adaptive landscape is that such a hyperspace has too many dimensions to be of any practical analytical value. One of the goals of theoretical morphospace construction, therefore, is to create geometric hyperspaces that contain realistic spectra of hypothetical morphologies but that at the same time possess a minimum number of dimensions.

In a thoughtful review, Hickman (1993b:172) lists what she considers to be seven misconceptions about how theoretical morphospaces are constructed:

(1) morphospaces cannot be created and explored without a computer; (2) morphospaces require simulation modeling; (3) morphospace must be "biologically correct'" (simulation parameters must be morphogenetic); (4) morphospace is multidimensional; (5) the axes of morphospace must be continuous variables; (6) there is one best coordinate system and set of axes for any given group of organisms; and (7) most groups of organisms are so complex that they defy parameterization of their basic geometry.

With this list Hickman (1993b) is attempting to strip away some of the mystique of theoretical morphospace, a mystique that has intimidated some from further exploring the concept and its applications. In this attempt she also is performing a valuable service to theoretical morphology. Of her list of misconceptions I would perhaps disagree with numbers two and four. One of the fundamental aspects of a theoretical morphospace is the creation of hypothetical form without any empirical measurement data from actual organic form itself. The creation of nonexistent form in particular usually requires some sort of simulation, if only for the purpose of visualization. But that simulation need not be at all complicated, as I shall demonstrate in a moment with a simple example. With respect to misconception number four, I think Hickman may mean that a theoretical morphospace need not *necessarily* be hyperdimensional, although most are. But minimizing the dimensionality of the morphospace required to examine a particular group of organisms adequately is indeed a major goal of theoretical morphology.

To emphasize the point that a theoretical morphospace does not have to be produced by a computer or to involve complex mathematics (although many do) I give a "theoretical morphospace of triangular form" in figure 2.4. This perfectly valid theoretical morphospace of triangular form was created with nothing more complex than a ruler, a pencil, a piece of paper, and an imagination. The simulations in the morphospace show the range of potential triangular form from tall thin triangles in the upper left to short broad triangles in the lower right. Although the triangles in themselves have no biological meaning, they could easily be used by a botanist to examine leaf shape in plants, by a vertebrate paleontologist to examine tooth shape in dinosaurs, or even by an anthropologist to examine arrowhead shape in neolithic human cultures. The last example is nonbiological in that a stone arrowhead is not, nor ever has been, alive. Yet the techniques of theoretical morphology might indeed be fruitfully used by anthropologists to simulate arrowhead morphologies that have never been created in any human culture and to ask the interesting question, "Why not?" which would lead immediately to interesting questions concerning arrow aerodynamics and structural limitations in using stone as an engineering medium. In fact, we shall encounter a "theoretical morphospace of conical form" designed for snails in chapter 4, one very similar to the theoretical morphospace of triangular form used here.

This simple theoretical morphospace of triangular form (figure 2.4) can even be used to illustrate some principles of growth, in that magnitude increases occur along both axes of the morphospace and magnitude increases generally occur in organic growth. Thus note that the triangles along the diagonal through the origin "grow" isometrically; that is, they become larger away from the origin, but they retain exactly the same shape. The triangles that "grow" parallel to either the x-or the y-axis in figure 2.4 grow anisometrically, in that their shape changes markedly with increasing size.

Of course, most theoretical morphospaces are more complex and elegant than the simple theoretical morphospace of triangular form given in figure 2.4. For example, many morphologists may wish to design a theoretical morphospace in which size is not a factor. In figure 2.4, a range of potential changes in triangular proportions are given, but the effect of absolute size is also present, as can be seen in the triangles along the diagonal. The proportions of these triangles are all the same; they differ only in size. The dimensional parameters of more

FIGURE 2.4. A theoretical morphospace of hypothetical "triangular form," created with nothing more complex than a ruler, a pencil, a piece of paper, and an imagination. Apex height and base width increase in equal increments for each simulated triangle.

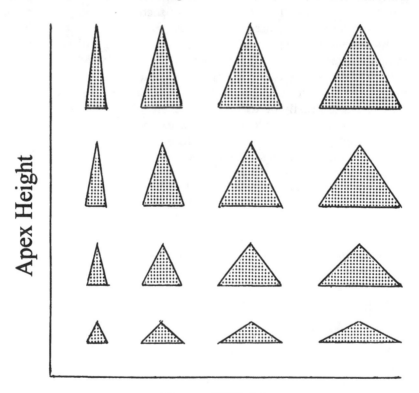

complicated theoretical morphospaces are sometimes chosen as ratios of measurements, rather than absolute measurements, to eliminate the direct effect of size and to produce "dimensionless dimensions."

The geometric model of form that was used to produce the theoretical morphospace in figure 2.4 is "isosceles triangle," a model with three straight lines whose interior intersecting angles always sum to 180° and with two sides of equal length. The parameters of the model used to define the dimensions of the theoretical morphospace are apex height and base width. The much more elegant geometric model of form used to produce the theoretical morphospace in figure 2.2 is "univalved shell form," a model involving rotations of a generating curve through space with the usage of logarithmic helicospirals. The parameters of the model used to define the dimensions of the theoretical morphospace are $W$ (whorl expansion rate) and $D$ (distance of the generating curve from the coiling axis).

Although the model parameters of the theoretical morphospace of triangular form are geometric, they also would be classified as biometric or morphometric if applied to real organisms. In Hickman 's earlier discussion (1993b), we noted that one of her list of common misconceptions concerning theoretical morphospaces was that the "morphospace must be 'biologically correct' (simulation parameters must be morphogenetic)." Although the parameters of a theoretical morphospace can be simple geometric measures (as in figure 2.4), many morphologists do prefer a theoretical morphospace in which the parameters are in some way morphogenetic. Such parameters are not simple static measurements of momentary form but are usually rate measurements of form change or growth.

Regardless of the complexity of the theoretical morphospace, it is important that the hypothetical forms in the morphospace simulate as closely as possible the actual organic forms under analysis. If the simulated forms are too generalized, then little can be learned from the theoretical morphospace concerning actual morphology. For example, it would be possible to create a theoretical morphospace of "boxes" by using the simple linear dimensions of length, width, and height. In this morphospace, a box whose length, width, and height are all the same is a cube. A box whose width and height are the same but whose length is much greater is a coffin. A box whose width and length are the same but whose height is much greater is a sarcophagus (i.e., a vertical coffin—ancient Egyptian burial containers are, for some unknown reason, most often displayed in this position in the museums of the world). Unfortunately for boxes, all kinds of things can be fit inside them (not just corpses); thus their potential explan-

**21**

atory (and theoretical exploratory) power is very small. The same simple height-width boxes can be used to explore the regions of morphospace occupied by amphineurans (Watters 1991), by the pygidia of trilobites (Labandeira and Hughes 1994), by the forms of egg cases of living and extinct chondrichthyans (McGhee 1982a, McGhee and Richardson 1982), and on and on. A box is a box, but a multiplated amphineuran skeleton is very different from a trilobite pygidium, and both are even more different from a shark egg case. Thus more complex morphospaces are necessary for a realistic theoretical morphologic analysis of these actual organic forms.

## THEORETICAL MORPHOSPACES VERSUS EMPIRICAL MORPHOSPACES

*For the morphometrician, morphospace is an arena for multidimensional analysis of actual measurements from real specimens. There is nothing theoretical about this kind of morphospace.*

*Hickman (1993b:172, emphasis Hickman's)*

The multidimensional spaces produced from the multivariate statistical analyses of actual measurement data are totally different from the measurement-independent theoretical morphospaces used in theoretical morphologic analyses. To clarify this difference, I have coined the term *empirical morphospace* and defined empirical morphospaces to be "multidimensional morphological spaces produced from the mathematical analysis of actual measurement data using such techniques as principal components analysis, factor analysis, Fourier analysis, or other polynomial series approximations of natural morphology" (McGhee 1991:96). Fractals (Mandelbrot 1983) could now also be added to this list, as fractal metrics have also been used to create empirical morphospaces (see Lutz and Boyajian 1995).

Empirical morphospaces can be very useful in characterizing the range of existent form in nature. Such morphospaces do have major limitations, however. The first major deficiency of an empirical morphospace is the sample-determined nature of its dimensionality. The dimensions of empirical morphospace are in fact determined by the very samples under analysis themselves, in which the "sample" in this case refers not only to the number of individual organisms in the analysis but also to the number of different morphologic characters measured from those organisms. In contrast, the dimensions of a theoretical morphospace are sample independent. Foote (1995), a morphometrician who has extensively used empirical morphospaces in his research, also clearly pointed out this problem in his study of the morphologic evolution of crinoids:

> The foregoing description of changes in morphological distribution illustrates one of the fundamental limitations of empirical, as opposed to theoretical morphospaces (Raup 1966, 1967; McGhee 1991). A large group of very similar species has the potential to "pull" the principal coordinates by virtue of its sheer numbers rather than its variability . . . also that the composition of the axes themselves can depend strongly on sampling. Recent work toward developing a more theoretical approach to crinoid morphology (Kendrick 1993) should help circumvent this limitation. (Foote 1995:288)

MacLeod (1996:254) also explained the "inherent indeterminacy" of empirical morphospace analyses, in that morphologic patterns observed in such spaces are "highly sensitive to the manner in which the underlying morphology has been sampled."

A second major deficiency of an empirical morphospace (and a much more crucial one, in my opinion) is that the dimensionality of empirical morphospaces is determined by existent form alone. The crucial ability of a theoretical morphospace to reveal nonexistent form is a function of the measurement-independent nature of its dimensions, dimensions that are determined by a geometric model, not by sample measurements. In contrast, an empirical morphospace can never reveal to us morphology that might have been or morphology that could be in the future. Foote (1995:280) also observed this deficiency was "a potential limitation with the foregoing analysis is that the principal-coordinate space does not completely encompass all the morphological variation among species. Moreover, the coordinate axes in no way reflect 'pure' form, but are influenced by the density with which various morphological themes are sampled (McGhee 1991)." Kendrick (1996:208) saw this same deficiency as well: "Morphospace analysis based on theoretical parameterization breaks down complex geometries into character sets that capture the breadth of morphological possibility. In contrast, empirical morphospaces based on specimen measurement or characters spotlight only the realized subset of potential morphology."

At first glance, empirical morphospaces can appear deceptively similar to theoretical morphospaces. Both are hyperdimensional. Existent morphology can be plotted in each, and frequency distributions of form can be contoured in each. For discussion and contrast, I include here an empirical morphospace (figure 2.5) from the very fine morphologic work of Swan and Saunders (1987). Their study of the morphology of Late Carboniferous ammonoids I consider to be an excellent example of a "combination study" (a term defined in the next section of this chapter), as those researchers used *both* empirical morphospaces *and* theoretical morphospaces in their study.

The main reason for choosing figure 2.5 as an example of an empirical morphospace is to contrast it with the theoretical morphospace in figures 2.2 and 2.3, since both morphospaces deal with ammonoids. Note that they look very similar, although their dimensions are radically different. The dimensions of the theoretical morphospace in figures 2.2 and 2.3 are $W$ and $D$, two geometric parameters of an abstract mathematical model of form. No measurements of real organic form are present in figure 2.2, yet the realistic-looking hypothetical ammonoids exist anyway, as they are model produced.

The dimensions of the empirical morphospace in figure 2.5 are "P.C. 1" and "P.C. 2." The "P.C." stands for "principal components," a multivariate statistical technique. The coordinates of the dimensions shown in figure 2.5 were determined by the principal-components analysis of a very large data matrix of measurements taken from Late Carboniferous ammonoids. The position coordinates of an actual ammonoid species in the empirical morphospace are loadings on the principal-components axes, a measure of the degree of contribution of that species to determining each axis. The dimensions of the empirical morphospace in figure 2.5 were determined by the ammonoid species themselves, and without any measurements of those species, the empirical morphospace would not exist.

The frequency distributions of actual ammonoid species in both the empirical morphospace (figure 2.5) and the theoretical morphospace (figure 2.3) have been contoured, and the two figures appear very similar in this respect. Figure 2.5 contains a total of 371 ammonoids, and figure 2.3 contains 405. Now consider what would happen if a 406th ammonoid were added to the theoretical morphospace in figure 2.3. The frequency distribution contours might change slightly to reflect the addition of this new morphologic datum, but the dimensions themselves of the theoretical morphospace would remain unchanged, and the positions in the morphospace of the previously plotted 405 ammonoids would remain unchanged. One could add an additional 1 million ammonoids to the theoretical morphospace, and the dimensions of the space would still remain unchanged, as would the positions in the morphospace of the original 405 ammonoids.

FIGURE 2.5. An empirical morphospace of ammonoid shell form, produced by the principal-components analysis of measurements taken from 371 specimens (representing 281 species) of Late Carboniferous ammonoids. Contours measure the density of specimens per unit area of the space produced by plotting the magnitudes of variation along principal-components axis 1 (P.C. 1) versus principal-components axis 2 (P.C. 2). The principal-components analysis of morphologic variation in Late Carboniferous ammonoids reveals eight morphotypes (roman numerals I through VIII) that correspond either to "topographic highs" or contour extrema in the empirical morphospace. *From Swan and Saunders 1987. Reprinted from* Paleobiology *and used with permission.*

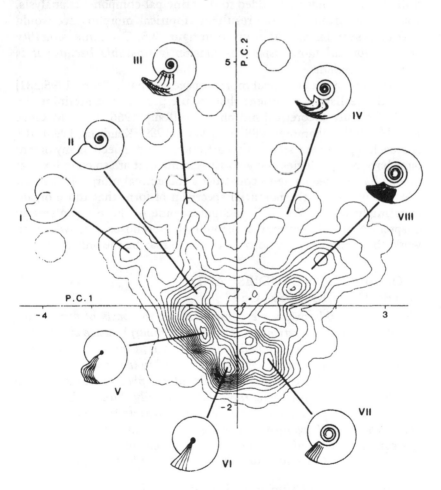

This is not true for the empirical morphospace. If measurements from a 372nd ammonoid were added to the original data matrix and the principal-components analysis were rerun, an empirical morphospace different from that shown in figure 2.5 would result. The very dimensions of the morphospace itself would have changed, and every position of the previously plotted ammonoids in that morphospace would also have changed. They might not have changed much, but they would have changed. And if measurements from an additional 1 million ammonoids were added to the principal-components analysis, you can be certain that the resultant empirical morphospace would bear no resemblance to that given in figure 2.5. *The dimensionality of an empirical morphospace is inherently unstable because it is sample dependent.*

In defense of the empirical morphospace approach, Foote (1995:281) made the realistic assessment that "while perhaps not attaining the ideal of a truly theoretical morphospace (Raup 1966, 1967; McGhee 1980[b], 1991; Okamoto 1988[a]; Ackerly 1989; Kendrick 1993), the approach represents a step in this direction." And indeed, many organisms have morphologies and growth systems that are so complex that no one has yet been able to construct a theoretical morphospace that adequately depicts the potential spectrum of form that those organisms might attain. We shall encounter some of those organisms in chapters 6 and 7. Without a theoretical morphospace in which to work, the empirical morphospace approach may be the only approach.

## "COMBINATION STUDIES" AND "HYBRID MORPHOSPACES"

*Theoretical morphospaces are constructed on the basis of a priori assumptions about theoretical morphology—the morphospace exists without the organisms being studied, which are then fit into it. . . . Empirical morphospaces, on the other hand, are developed in the opposite direction. Large numbers of organisms are analyzed morphometrically and the morphospace developed using these data, typically through the application of ordination methods. . . . We will demonstrate that, for the study of a single group of organisms, there is great utility in developing morphospace models that conform to both the theoretical and empirical models.*
*Chapman, Rasskin-Gutman, and Weishampel (1996:66)*

Chapman, Rasskin-Gutman, and Weishampel (1996) make an excellent point. Theoretical morphospaces and empirical morphospaces are very different, so why not use them both? There are indeed studies that have used both theoretical and empirical morphospaces in the analysis of form in nature, and these works I have termed "combination studies," as they combine both theoretical and empirical analytic techniques. An early example of a combination study is Williamson's (1981) analysis of the evolution of gastropod form in the Turkana Basin. Williamson used empirical morphospace techniques to analyze the stratigraphic aspect of observed evolutionary changes in gastropod form, and theoretical morphospace techniques to analyze the developmental aspects of those morphologic changes. Another excellent example are the studies by Saunders and Swan (1984), Swan and Saunders (1987), and Saunders and Work (1996) of the morphology of Late Carboniferous ammonoids. Those works use theoretical morphospaces for aspects of ammonoid morphology in which theoretical morphospaces exist (shell form), and empirical morphospaces for aspects of ammonoid morphology in which no theoretical morphospace has as yet been designed (shell ornamentation).

Another conceptual possibility is to try to combine the two analytic techniques, to create a morphospace that has both empirical and theoretical aspects. Such morphospaces do exist, and I have tentatively termed them "hybrid morphospaces," as they are partially dependent on measurements taken from actual organisms yet at the same time are produced by theoretical morphologic techniques of form simulation. An early type of hybrid morphospace can be seen in Waters's (1977) analysis of pentremites calyx morphology (figure 2.6). Waters took measurements from the marginal outline of 30 pentremites individuals. Measurements for each individual were fit to a Fourier series, and then an average Fourier series for all 30 individuals was calculated. The harmonics of that average Fourier series were then systematically varied to produce the simulated pentremites calyx outlines given in figure 2.6.

In a somewhat different type of analysis, Spivey (1988) constructed an empirical morphospace of shell shape in barnacles. He then used the clusters of morphologies found in the empirical morphospace to define four morphotypes of barnacle shell form: (1) paraboloid, (2) conical, (3) frustrum of a cone, and (3) frustrum of an ellipsoidal cone. The mean morphometric measures of these four morphotypes were used to produce computer simulations of hypothetical barnacle shell form. Those hypothetical simulations demonstrated that species with conical or frustrum of an ellipsoidal cone geometries have smaller shell volumes than do species with paraboloid or frustrum of a cone geometries, even though the same species appear not to exhibit a relation-

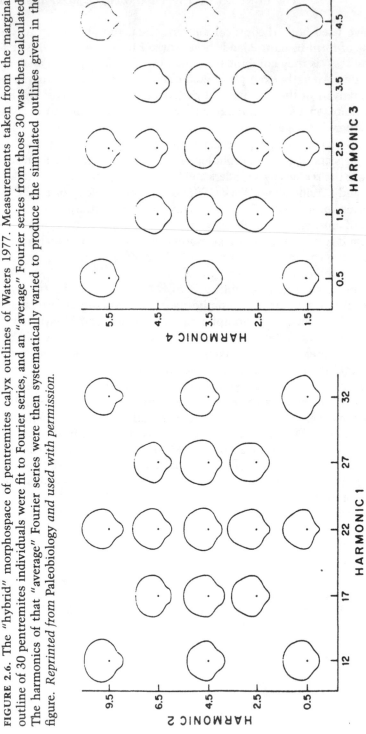

FIGURE 2.6. The "hybrid" morphospace of pentremites calyx outlines of Waters 1977. Measurements taken from the marginal outline of 30 pentremites individuals were fit to Fourier series, and an "average" Fourier series from those 30 was then calculated. The harmonics of that "average" Fourier series were then systematically varied to produce the simulated outlines given in the figure. *Reprinted from Paleobiology and used with permission.*

ship between the absolute size of the shell and its geometry. Although Spivey did not actually use these simulations to construct a hybrid morphospace of barnacle shell form, such a space could easily be created.

A final elegant example of a hybrid morphospace is given in figure 2.7. Stone (1996a) took measurements from actual *Cerion* specimens and then statistically fit those measurements to allometric growth equations. He systematically varied the parameters of the equations by using the computer program "CerioShell" (Stone 1995) to produce the simulated shell height versus shell width combinations seen in the *Cerion* shells given in figure 2.7.

## THEORETICAL MORPHOSPACES VERSUS THEORETICAL DESIGN SPACES

*Theoretical design space . . . subsumes theoretical morphospace.*
*Hickman (1993b:180)*

Theoretical design spaces are theoretical hyperspaces that include dimensions that are not morphologic. As defined by Hickman (1993b: 170), a theoretical design space "differs from a morphospace by combining morphological axes with axes that are ecological, behavioral, or physiological."

At the present time, the number of examples of theoretical design spaces are few. Hickman (1993b) herself parameterized "wheel space," a theoretical design space of rotating locomotory structures for organisms, and "suspension feeding space" for trochoidean gastropods. The most ambitious and comprehensive theoretical design space yet to be created is the "skeleton space" of Thomas and Reif (1991, 1993), a theoretical hyperspace of possible structural element combinations in animal skeletons. The space is seven dimensional, with each dimension specifying a character state that may exist in two to four different conditions (figure 2.8).

The number of potential combinations of skeletal character conditions in the skeleton space is 1,536, but Thomas and Reif (1993) reduced this large number to 186 by considering the skeletal character conditions in a pairwise fashion. Interestingly, they argued that eight of the potential 186 character condition pairs were "illogical or inviable" combinations; thus the skeleton space succeeded in specifying impossible skeletal characteristics, a capability characteristic of theoretical morphospaces. Analysis of actual organic skeletons reveals that nearly the entire skeleton space is filled, with only five of the 178

**FIGURE 2.7.** The height-width "hybrid" morphospace of *Cerion* gastropod morphology in Stone (1996a). Measurements taken from actual *Cerion* specimens were statistically fit to allometric growth equations, and then the parameters of the equations were systematically varied to produce the simulated shell height versus shell width combinations seen in the *Cerion* shells given in the figure. All the simulated *Cerion* shells are standardized to the same height in the figure to accentuate the shape differences produced in the simulations. *Artwork courtesy of J. R. Stone.*

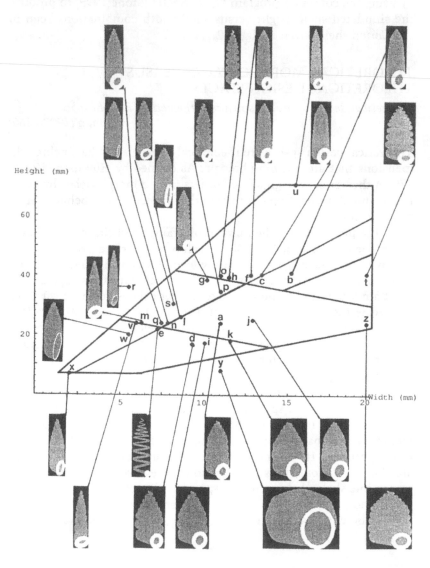

**FIGURE 2.8.** Illustrations of the seven dimensions of Thomas and Reif's skeletal theoretical design space (1993). Dimension 1 is Topology, which may be either internal (2.8A) or external (2.8B). Dimension 2 is Material, which may be either rigid (2.8C) or flexible (2.8D). Dimension 3 is Number, which may be one element (2.8T), two elements (2.8V), or three or more elements (2.8W). Dimension 4 is Geometry, which may be rods (2.8G), plates (2.8H), cones (2.8J), or solids (2.8K). Dimension 5 is Growth Pattern, which may be accretionary (2.8L), unit/serial (2.8M), replacement/molting (2.8Z), or remodeling (2.8N). Dimension 6 is Building Site, which may be either in place (2.8X) or in prefabrication (2.8Y). Dimension 7 is Conjunction, which may be in no contact (2.8P), jointed (2.8Q), sutured/fused (2.8R), or imbricate (2.8S). *Artwork courtesy of R. D. K. Thomas.*

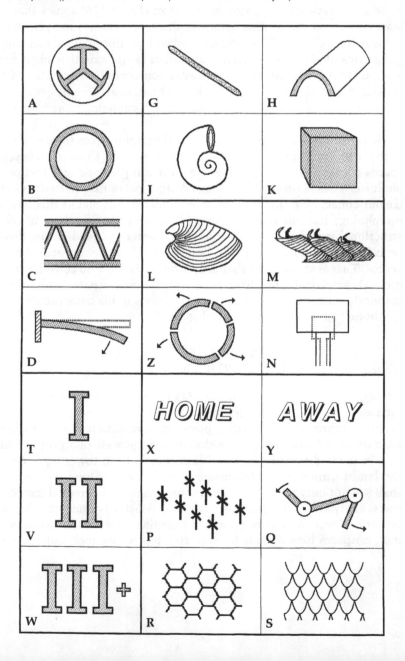

possible character combinations apparently not found in nature (Thomas and Reif 1993:fig. 12). Thomas and Reif (1993:341) concluded their study by contending that the "organizational properties of animal skeletons suggest that their design elements are fixed point attractors, structures that we characterize as topological attractors that evolution cannot avoid."

Although Thomas and Reif (1991, 1993) consider the skeleton space to be a theoretical morphospace, Hickman (1993b:174) argues that at least two of the seven dimensions of the space are not morphologic: "One is the axis that differentiates rigid and flexible skeletal elements (properties of skeletal materials), and other is the axis that identifies a set of four pattern-forming processes (ontogeny)." Thus Hickman considers the skeleton space to be a theoretical design space and praises it as a "giant step in the direction of revitalizing morphospace" (Hickman 1993b:173).

Hickman (1993b) considers theoretical morphospaces to be a subset of the larger concept of a theoretical design space, as theoretical design spaces contain dimensions that are both morphologic and nonmorphologic. The creation of theoretical design spaces may indeed revitalize our concepts of the analysis of morphology. But just as theoretical morphology has not been subsumed by the synthetic school of constructional morphology (as discussed in chapter 1), I predict that theoretical morphospaces will not be subsumed by theoretical design spaces. That assessment is shared by Stone (1997b), who considers theoretical, empirical, and theoretical design morphospaces to be three distinctly different and coexisting dimensions in his construction of a synthetic "space of morphological spaces."

## THE CURRENT STATE OF THEORETICAL MORPHOSPACE CONSTRUCTION

Enough of concepts and definitions! Let us now jump into the hyperdimensional vistas of form seen in actual theoretical morphospaces. In chapters 3 through 7, we shall explore the distribution of both realized and unrealized morphologies produced by the process of organic evolution in many different forms of life on Earth, from towering trees on the land to microscopic foraminifera in the seas. Along the way we shall see not only what has been accomplished in theoretical morphospace analyses in the past 30 years but also how much remains to be done. There are entire groups of organisms for which theoretical morphospaces have already been created but whose morphologic evolution has not been examined, simply because no one has bothered yet to examine them using the analytic techniques of theoretical morphology. Graduate students considering research in the field of morphology should particularly note that challenging potential Ph.D. dissertation topics abound in the chapters to come.

# III

# Twists and Twigs: Theoretical Morphospaces of Branching Growth Systems

*In matters of visual form we sense that nature plays favorites.*
*Among her darlings are spirals, meanders, branching patterns, and*
*120-degree joints. Those patterns occur again and again. Nature*
*acts like a theatrical producer who brings on the same players each*
*night in different costumes for different roles. . . . A look behind the*
*footlights reveals that nature has no choice in the assignment of*
*roles of players. Her productions are shoestring operations,*
*encumbered by the constraints of three-dimensional space.*
*—Stevens (1974:3–4)*

In this and the next three chapters we shall examine the range of potential form available to nature for its organic players, as revealed by
the creation of theoretical morphospaces of hypothetical form. In this
chapter we shall examine theoretical morphospaces created for organisms, both plant and animal, that use branching growth systems. In
chapters 4 and 5 we shall examine the various theoretical morphospaces that have been designed for organisms using accretionary
growth systems, and in chapter 6 we shall examine theoretical morphospaces created for organisms using discrete growth systems.

Much of morphology in nature may be simulated by tubular forms
that do not simply expand or elongate but, rather, bifurcate, trifurcate,
and so on. Of interest is the fact that organisms that grow via a process
of branching are common in both the plant and animal kingdoms. Although plants and animals are very different biologically (plants are
photoautotrophs and animals are not), they both often grow in a fashion to maximize the surface area of their branching geometries. On
the other hand, many of the geometric constraints found in both animals and plants are associated with the maintenance of branch avoidance (except in those organisms in which subsequent branches may
anastomose) and the structural limitations of branch strength. A great

deal of attention has been given to producing theoretical models of the morphogenesis of branching form in plants and animals, and a selection of some these models will be considered in greater detail in chapter 10.

Much less attention has been given to the creation of theoretical morphospaces of branching form, which is surprising considering the large number of theoretical models of morphogenesis for this same type of organic form in nature. Those studies that have followed, or at least considered, the theoretical morphospace methodology will be considered in this chapter. The theoretical morphospaces of hypothetical branching morphologies that do exist have been used to explore the range of form that could be produced in such disparate organisms as bryozoans, graptolites, and trees. Parameterization of the geometry of branching growth is usually more complex than that of non-branching morphology, with the subsequent morphospaces produced being of a higher dimensionality.

## THEORETICAL MORPHOSPACES OF BRANCHING FORM IN MARINE ANIMALS

Van Valen (1978) designated as "arborescent animals" a diverse and phylogenetically disparate grouping of marine animals that have the form of miniature trees. The Porifera and Cnidaria often produce laminar morphologies (covered in chapter 6), but these phyla also produce arborescent animal forms, particularly the Cnidaria (hydrozoans like *Millepora* and anthozoan octocorals like *Gorgonia*). Arborescent animal form is the rule rather than the exception in the phylum Bryozoa, in the class Crinoidea in the phylum Echinodermata, and in the classes Graptolithina and Pterobranchia in the phylum Hemichordata. The morphologies of crinoids are strikingly like tropical trees in their divisions of rootlike holdfasts, trunklike stems, and disklike canopy of branched arms. The many planktic forms of graptolites are particularly interesting in that their arborescent forms grow downward rather than upward; that is, their treelike morphologies hang down from a colonial swim bladder or other floating organ rather than branching up from the seafloor.

The simplest way to model branching form in nature is to begin with branching in a two-dimensional plane. In that two-dimensional plane, the simplest type of branching is bifurcation, that is, one structural element (a branch, twig, stem, and so on) that simply bifurcates, producing two new structural elements. These two structural elements themselves then simply bifurcate, producing four new struc-

**35**

tural elements and a total of seven structural elements in the system. The number of growing tips in such a simple bifurcation system thus increases in a base 2 geometric series:

$$\text{number of growing tips} = 2^0 + 2^1 + 2^2 + 2^3 + \cdots + 2^n \quad (3.1)$$

At this point we can introduce our first model parameter with a question: what is the angle between the structural elements in the bifurcation? The simplest bifurcation model is one in which the bifurcation angle remains constant over time. A more complex (and realistic) model allows the bifurcation angle to vary over time.

Given the bifurcation angle as one parameter, a second model parameter can also be introduced with a question: when does the structural element bifurcate? The simplest bifurcation model is one that bifurcates whenever a given structural element reaches a certain length during growth. In such a bifurcating system, all the structural elements are exactly the same length. A more complex (and realistic) model allows the lengths of the structural elements to vary over time.

These two parameters are sufficient to begin to examine some patterns of branching form in nature, such as encrusting bryozoans. As encrusters, their colony form may be modeled as if it were a two-dimensional branching system. Gardiner and Taylor (1980, 1982) studied 17 colonies of two species of the Jurassic cyclostome bryozoan *Stomatopora* by examining the "angle of branch dichotomy" (which they termed $\theta$) and the "distance between successive branching points" (which they termed $l$). The parameter $l$ is measured by the number of zooids comprising the branches between bifurcation points. Interestingly, they found that the lengths of the structural elements ($l$) in *Stomatopora* colonies varied little, whereas there was a great deal of variation in the bifurcation angles between the structural elements ($\theta$) in the colony.

Why do the bifurcation angles vary within encrusting bryozoan colonies? To address that question, Gardiner and Taylor (1980, 1982) began to computer-simulate hypothetical bryozoan colonies. Each simulated colony branch starts from a single point (like the founder zooid of an actual colony), growing in length by terminal addition (like the budding of zooids at branch tips) and periodically bifurcating (as uniserial rows of zooids do). They then added an important morphogenetic constraint to the model: a growing branch terminates growth whenever it intersects another, previously established, branch. The parameter $l$ was set at an average of two zooids per internode. A series of simulations were run in which (1) the bifurcation angle ($\theta$) was held constant throughout all bifurcations during colony growth, in which (2) the bifurcation angle was progressively decreased by a fixed amount in each successive generation of bifurcations, in which (3) the bifurcation angle was increased by a fixed amount in each generation, in which (4) the bifurcation angle was arithmetically decreased each generation until a minimum value was reached, in which (5) the bifurcation angle was arithmetically increased each generation to a maximum and, last, in which (6) the bifucation angle was decreased exponentially to a minimum. They also added additional complexity to the simulations in modeling the changes in bifurcation angle between generations (simulation models 2 through 6). In a given generation, all the bifurcation angles could be the same (but different from those of the previous and subsequent generations), or they could be chosen from a distribution of angles with a given mean value and standard deviation for each generation.

From all their computer simulations of hypothetical encrusting bryozoan colonies, Gardiner and Taylor (1980, 1982) found that the simulations in which the bifurcation angle decreased exponentially each generation, and in which the value of the bifurcation angle in each generation was allowed to vary about a mean value, best matched the observed patterns of colony form seen in *Stomatopora*. They concluded that such a growth pattern was optimal in producing "a colony of roughly circular shape with radial growth from a maximum number of surviving growing points" and that such a morphology "gives even and economical coverage of substrate and also has high survival potential if part of the colony encounters unfavourable substrate and ceases to grow" (Gardiner and Taylor 1980:107).

Gardiner and Taylor (1980, 1982) did not explicitly construct a theoretical morphospace in their study, but they did follow the "theoretical morphospace methodology" in that they created a geometric continuum of hypothetical colony forms, colonies that are mathematical constructs having no real biological existence. They then compared actual bryozoan colonies with the spectrum of form seen in the hypothetical colonies. They also completed the final important step in theoretical morphospace methodology (see chapter 2) by analyzing the functional significance of the match seen between real colony form and one of their many hypothetical colony forms. Here the methodology of theoretical morphology also helped in the functional analyses by producing both existent and nonexistent morphology. That is, Gardiner and Taylor (1980, 1982) could not ask only, "Why does *Stomatopora* colony form match the geometry seen in this computer simulation?" they could also ask, "Why does *Stomatopora* colony form *not match* the geometry seen in these other, entirely possible, computer simulations?"

A similar "theoretical morphospace methodology" analysis of two-dimensional colony form was conducted by Fortey and Bell (1987) for organisms very different from the benthic-encrusting colonies of bryozoans examined in the previous study. Instead, Fortey and Bell (1987) looked at the branching geometries found in planktic colonies of multiramous graptoloids, colonies that passively float in the waters of the ancient Paleozoic seas. Because the ecologies of benthic and planktic animals are vastly different, from the beginning one might expect that benthic and planktic colony forms would also be vastly different. As we shall see, this was not the case.

To examine graptoloid colony form, several new complexities must be added to our previous consideration of branching possibilities in a two-dimensional plane. In the earlier discussion, a single structural element was the starting point. Graptoloids may have a single structural element (the stipe) as the starting configuration, but they may also have two (biradiate origin) or three (triradiate origin) or four (quadriradiate origin). In the previous discussion, the structural elements always bifurcated, with an initial or primary structural element producing two new or secondary structural elements. In a bifurcation process, the angle between the two secondary structural elements is bisected by the axis of the primary structural element; that is, in the bifurcation process, one secondary structural element diverges at an angle $A$ to the left of the line of axis of the primary structural element, and the other secondary structural element diverges at the same angle $A$ to the right of the line of axis of the primary structural element, and the bifucation angle between the secondary structural elements is simply twice the magnitude of the angle $A$. Alternatively, the primary structural element may branch in a dichotomy in which the angle of the secondary structural element to the left of the axis of the primary structural element is $A$, and the angle of the secondary structural element to the right is $B$, with the result that the branching angle between the two secondary structural elements is not bisected by the line of the axis of the primary structural element. In addition, in graptoloid colonies, there may be a "main stipe" that does not branch at all but simply elongates by the addition of new thecae while "subsidiary stipes" branch off at angles along the main stipe. Finally, in the previous discussion, the structural elements of the colony always grew in straight lines. In graptoloid colonies, the stipes may grow in straight lines, but they may also curve.

The mathematical simulation of a floating graptoloid colony is thus more complex than the simulation of an encrusting bryozoan. To produce hypothetical graptoloid colonies, Fortey and Bell (1987) actually used a computer program initially designed to simulate rhizomatous growth in plants (Bell, Roberts, and Smith 1979). Similar to Gardiner and Taylor's bryozoan branching model (1980, 1982) are the features that graptoloid stipes grow at their tips only and that this growth is terminated if a growing stipe encounters an already existent stipe. The latter growth constraint produces colony forms that remain two dimensional. Such a constraint for a planktic colony, freely floating in the three-dimensional water column, may seem at first unrealistic, unlike for an encrusting bryozoan colony growing over a benthic substrate surface. Examination of actual graptoloids reveals, however, that they do indeed maintain two-dimensional rhabdosomes. Fortey and Bell (1987) did modify their growth algorithms to produce graptolite colonies with stipes arranged in three dimensions, but in doing so they created nonexistent graptoloids.

As mentioned earlier, Fortey and Bell (1987) did not explicitly create a theoretical morphospace of graptoloid colony form. Like Gardiner and Taylor (1980, 1982), however, they did follow the "theoretical morphospace methodology" in that they created a multiplicity of hypothetical colony forms, colonies that are mathematical constructs having no real biological existence, and then compared actual graptoloid colonies with the spectrum of form seen in the hypothetical colonies. They concluded that their computer simulations "with close natural analogues are those which achieve wide spread of the planar colony while avoiding overlap between branches" leading to the "efficient and regular distribution of zooids through the area included by an essentially planar rhabdosome" (Fortey and Bell 1987:1, 9).

Of particular interest is an observed evolutionary trend in the reduction of "primary stipes" in Ordovician graptolites with dichotomous branching. These graptolites begin with colony forms with quadriradiate origins, which are reduced to triradiate, to biradiate, and ultimately to uniradiate in descendant lineages. With computer simulation, it was possible to show that each reduction in the number of primary stipes allows the progressive increase in the amount of area enclosed by the rhabdosome. Fortey and Bell (1987:14) concluded that "the reduction of primary stipes in Ordovician graptolites with dichotomous branching may be explained by the simpler acquisition of larger colonies with a lesser chance of interference between stipes. The selective pressure favoring larger colonies may have been toward in-

39

creased fecundity." Fecundity of the colony is a function of the number of zooids included in the rhabdosome. Thus two-dimensional colonies of both benthic bryozoans and planktic graptoloids are geometrically similar, even though their particular branching growth patterns are quite different. Both groups of organisms converge on branching geometries that maximize the number of zooids enclosed in the surface area of the colony.

In all the graptoloid colonies studied by Fortey and Bell (1987), the rhabdosome is a two-dimensional surface, and this surface itself is either strictly planar (i.e., it is flat and two dimensional) or gently conical (it is curved into the third dimension). The reader can easily visualize the latter geometry as a pancake that someone has draped over a globe of the Earth. The pancake remains a "two-dimensional" circular object yet is deformed into three dimensions in clinging to the surface of the globe, which is a three-dimensional sphere. McKinney (1981) studied the general geometry of "planar" colonies that may be complexly folded into the third dimension, as these geometries are common in sievelike or meshlike colonies of animals that form filtration sheets. McKinney termed such colony forms "colonial fans." They are common in gorgonacean cnidarians and fenestrate bryozoans. In these organisms, the branches of the colony are often connected to one another by lateral linkages, smaller structural elements that extend out from the side of one branch and connect to the side of an adjacent branch. Sometimes these lateral linkages are produced by anastomosis of the branches themselves. That is, a structural element bifurcates into two new structural elements, which then fuse back into a single element at some later time and which then rebifurcate in further growth.

McKinney (1981) noted that two common branching characteristics are found in colonies that form planar filtration sheets: (1) the spacing between the branches in the colony is very regular and uniform throughout the colony surface, and (2) the branches continue all the way to the periphery of the colony from their point of origin in the colony. The latter characteristic is quite unlike that seen in the branching geometries of the encrusting bryozoan colonies or floating graptoloid colonies considered previously, both of which generally terminate branches within the colony in order to avoid branch overlap (see Fortey and Bell 1987; Gardiner and Taylor 1980, 1982). Similar to both encrusting bryozoans and floating graptoloids, however, the "filtration sheet" organisms maximize surface area of the colony within the constraints imposed by the geometry of their branching system. Their particular geometries are constrained to produce a surface uniformly filled with organisms that are more or less equidistant from one another and that filter water that flows through the colony meshwork.

In theoretical morphology, the simplest form of filtration sheet "colonial fan" is an erect and inverted equilateral triangle in which the apex of the triangle is attached to the seafloor. This triangular or fan-shaped form is produced by a series of branches, which bifurcate at regular intervals and continue to grow upward until the next bifurcation, all the while with each branch maintaining a more or less constant distance from branches on either side. The degree to which the triangular fan flares or "opens" upward is a function of the frequency of bifurcation of the branches, as the spacing between the branches remains uniform throughout the colony.

In addition, the simplest triangular "colonial fan" is a flat filtration sheet that lies in a two-dimensional plane. The lateral triangular edges of this sheet often may be bent forward around the vertical axis of the colony to produce parabolic or half-conical three-dimensional forms. If the edges are bent forward and around enough to meet one another, a complete hollow cone results. Alternatively, the sheet may be complexly folded in an irregular pattern without a distinct geometry. Regardless of the complexities introduced by folding of the filtration sheet, the colony remains planar in that the sheet itself is produced by a two-dimensional branching process.

The most beautiful and complexly three-dimensional forms produced by folded "colonial fans" are those found in the spiral colonies of the extinct fenestrate bryozoan *Archimedes*. The theoretical morphology of helicospirally folded colonial fans was the focus of a second study by McKinney and Raup (1982). They also created an explicit theoretical morphospace of helicospiral colonial-fan form, one in which the geometric limits and hypothetical variations in this type of morphology could be explored.

The theoretical model of morphogenesis for helicospiral colonial-fan form that McKinney and Raup (1982) developed contains seven growth rules, or variables, necessary for the computer simulation of hypothetical colony forms produced by conically folding a planar surface (formed by bifurcating branches) around a central axis and then allowing the folded planar surface to climb the central axis along a helix. They used three of these variables to define the three dimensions of their theoretical morphospace:

1. *BWANG* = "angle between branches and axis of central helix."
2. *ELEV* = "rate of climb of helix."

**41**

3. $XMIN$ = "minimum distance between three adjacent branch tips at which the central branch bifurcates."

A two-dimensional slice through the three-dimensional theoretical morphospace of helicospiral colonial-fan form is given in figure 3.1, illustrating the morphologic effects of varying the values of the parameters $BWANG$ and $ELEV$. In helicospiral colonial fans, the bifurcating branches that form the filtration sheet originate from a central helical axis. Theoretically, the angle at which they intersect the axis ($BWANG$) may vary between 0° and 90°. At 90°, the branches stick straight out from the central axis (hence the planar surface of the filtration sheet does as well), and the colony has a morphology resembling that of an auger or woodscrew (see the computer simulation in the bottom right corner of figure 3.1). With decreasing values of $BWANG$, the filtration sheet is folded closer and closer to the central helical axis (see the simulations in the bottom center and bottom left corner of figure 3.1). The parameter $ELEV$ measures the degree of compression or expansion of the windings of the helix. At very low values, the windings of the helix are very close to one another (hence the surfaces of the filtration sheet between successive windings are as well, as seen in the simulation in the bottom right corner of figure 3.1). Increasing the value of $ELEV$ progressively stretches the windings of the helix in the vertical dimension, with the result that the successive surfaces of the filtration sheet are farther and farther apart (see the simulations, from bottom to top, along the right margin of figure 3.1).

The effects of varying the value of the parameter $XMIN$ is illustrated in figure 3.2, in which a two-dimensional slice has been taken through the theoretical morphospace at the value of 60° on the $BWANG$ axis. The parameter $XMIN$ essentially controls the spacing of the bifurcating branches in the filtration sheet folded about the helical axis. At low values of $XMIN$, the density of branches in the planar filtrations sheet is high (see the simulations along the left margin of figure 3.2). Increasing the value of $XMIN$ reduces the density of branches in the filtration sheet, as can be seen by comparing the computer simulations on the left margin of figure 3.2 with those on the right margin of the figure.

McKinney and Raup (1982) take the next step in theoretical morphologic analysis by plotting the position of two existent bryozoan species in their morphospace of hypothetical colonial form, that of the Paleozoic fenestrate species *Archimedes intermedius* and the modern cheilostome species *Bugula turrita* (points A and B, respectively, in

**FIGURE 3.1.** A sectional slice through the three-dimensional theoretical morphospace of helicospiral colonial-fan morphologies of McKinney and Raup (1982), illustrating the range in form produced by varying the parameters *BWANG* and *ELEV*, where the parameter *XMIN* is held constant at 30. The position of the Paleozoic species *Archimedes intermedius* and the modern species *Bugula turrita*, respectively, are indicated by points A and B in the morphospace. *From McKinney and Raup 1982. Reprinted from* Paleobiology *and used with permission.*

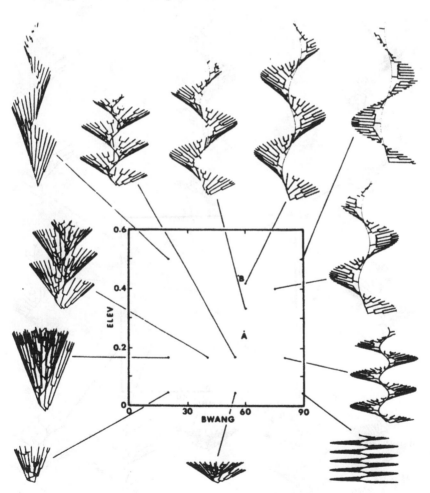

FIGURE 3.2. Another section through the theoretical morphospace of heli-cospiral colonial-fan morphologies, illustrating the range in form produced by varying the parameters *XMIN* and *ELEV*, where the parameter *BWANG* is held constant at 60°. The location in the morphospace of *Archimedes intermedius* and *Bugula turrita* indicated by points A and B. *From McKin-ney and Raup 1982. Reprinted from* Paleobiology *and used with per-mission.*

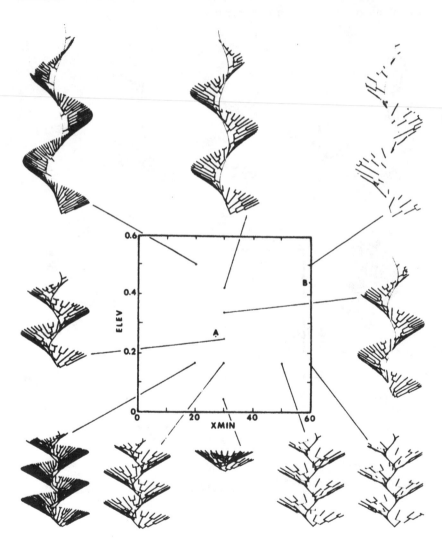

figures 3.1 and 3.2). The final step in theoretical morphologic analysis, exploring the functional significance of regions in the morphospace, is only briefly touched on in McKinney and Raup. Why does *Archimedes intermedius* occur in one region and *Bugula turrita* in another, in figures 3.1 and 3.2? They offer a brief discussion of the potential functional significance of spiral form in filter feeding organisms, but with respect to the theoretical morphospace itself, they simply state that "the functional effects of variation of the various growth parameters are essentially unknown" (McKinney and Raup 1982:110).

It is clear that the helicospiral colonial-fan geometry offers an extremely large surface area of filtration sheet relative to both the area of the point of attachment of the colony helix axis and the area of seafloor shaded by the colony (McKinney 1981, McKinney and Raup 1982). Minimizing the degree of colony contact with the substratum is potentially advantageous, given the greater probability of disturbance of the colony filtration surface by other benthic organisms on the sea bottom. Alternatively, potential attachment sites on the substratum may be rare or small; thus the erect helicospiral colony form again offers the attainment of a very large number of zooids in only a few colony structures. The obvious next step in the theoretical morphologic analysis of helicospiral colony fans is measuring and plotting the morphospace positions of additional real species in the morphospace. Which potential colony forms have been produced in nature? Which have not, although geometrically possible? And why?

Equally obvious is the exploration of some of the geometric properties of the morphospace itself. What are the magnitudes of filtration sheet surface area developed in each of the different computer simulations given in figures 3.1 and 3.2? It should be possible to map regions of differing surface area in the theoretical morphospace based simply on the differing geometries of colony form produced by variation in the morphospace's parameter values.

A major theoretical morphospace analysis of "meshwork" colony form in fenestrate bryozoans was conducted by Starcher (1987). He created a seven-dimensional theoretical morphospace of meshwork colony form and then examined the distribution of 128 species of fenestellid bryozoans in that morphospace. Starcher argued that the relative displacement and distribution of fenestrate higher taxa in the theoretical morphospace could be seen as a "canalization" of form resulting from the number of zooid rows in the branches of the colony.

The conceptual shift from two- to three-dimensional branching form was made in an interesting series of papers by Cheetham, Hayek, and Thomsen (1980, 1981) and Cheetham and Hayek (1983). In addition, an explicit theoretical morphospace of three-dimensional branching form was created by Cheetham and Hayek (1983). These morphologists were interested in the theoretical morphology of bryozoan colony form (like Gardiner and Taylor 1980, 1982; McKinney 1981, McKinney and Raup 1982) but chose to examine the morphology of three-dimensional arborescent bryozoan colonies rather than two-dimensional encrusting colonies or planar colonial fans.

Cheetham, Hayek, and Thomsen's original model (1980) is a two-dimensional one and treats arborescent branching as if it occurs only in a plane, like the branching pattern seen in river networks rather than three-dimensional trees. The model also simulates only a process of repeated bifurcation and lengthening of branches and thus is similar to that of Gardiner and Taylor (1980, 1982). In Gardiner and Taylor's simulations, however, branch lengths are not variable, and branches in different regions of the colony simulation are designated in terms of "generations" of successive bifurcations after the initial bifurcation. In Cheetham, Hayek, and Thomsen's model branch lengths are a major variable, and branches are classified as to "order," that is, first order, second order, and so on, as is done in some analyses of tree- and river-branching systems (Horton 1945, Strahler 1952). A first-order branch is one at the very tip of the branching "tree," distalmost from the "trunk" of the tree, and contains the growing tip itself. A second-order branch is one that contains two first-order branches extending from a bifurcation, and so on. The length of branch extending from one bifurcation point to another (which I have been terming a "structural element" in general) is specifically designated as a "link" in Cheetham, Hayek, and Thomsen's model.

The number of growing tips in the model follows a generalized mathematical series, the nature of which is determined by three parameters:

1. $R_l$ = the "link-length ratio," the ratio between the lengths of links originating at the same bifurcation.
2. $R_g$ = the "relative growth ratio," the ratio between the lengths of first-order branches originating at the same bifurcation.
3. $R_B$ = the "branching ratio," the ratio between the number of first- and second-order branches (or between numbers of branches belonging to different orders in general).

Using these parameters, Cheetham, Hayek, and Thomsen (1980) defined three classes of growth models: unequal branch growth ($R_g$

greater than 1.0), where the increase in the number of growing tips (first-order branches) is determined by the product of $R_g$ and $R_l$ ("Model I"); unequal branch growth ($R_g$ greater than 1.0), where the increase in the number of growing tips is determined by the quotient of $R_g$ and $R_l$ ("Model II"); and equal branch growth ($R_g$ equal to 1.0), ("Model III"). In Model I, bifurcation occurs more frequently in branches that grow faster, and in Model II, the reverse is true. These differences collapse in Model III, where branches grow at an equal pace. In a detailed statistical analysis of measurements taken from eight modern species of "adeoniform" (erect branching) cheilostome bryozoans, Cheetham, Hayek, and Thomsen (1980) found that growth Model III did not exist in the data. Only one species (*Cystisella saccata*) had a $R_g$ value that might not be significantly different from one. Of the other species, four fit Model I and three fit Model II.

Perhaps more interesting (from a theoretical morphospace rather than theoretical morphogenetic viewpoint) was that they found very low values of $R_B$ (2.04 to 2.51) in their actual bryozoans, values much smaller than those generally found in trees and close to the $R_B$ of "symmetrical" branching (2.0). Unlike trees, where a distal portion of branches may be more or less a miniature of the branching pattern of the entire tree, they concluded that "rather than being a miniature, a distal piece of a bryozoan is a more or less close representation of the whole branching structure at a less-developed stage of growth" (Cheetham, Hayek, and Thomsen 1980:361). That conclusion further suggests a high degree of direct internal control on the branching process in arborescent bryozoans.

Cheetham, Hayek, and Thomsen's original model (1980) considered branch elongation and bifurcation in a two-dimensional plane. Branch widening and thickening, branch strength, and bifurcation in three dimensions were added to the original model by Cheetham, Hayek, and Thomsen (1981), Cheetham and Thomsen (1981), and Cheetham and Hayek (1983). Arborescent bryozoan branches are different from trees, where a branch is more or less cylindrical with a circular cross section. The cross section of a bryozoan branch is a flattened ellipse, with a major and minor axis; hence the branch has a width and a thickness. The width of the branch depends on the number of rows of zooids present. The thickness of the branch is initially two zooids (back to back, facing out) but becomes thicker as additional skeletal material is added by those zooids with time.

Cheetham and Hayek (1983) made computer simulations of branching in three dimensions by adding two additional parameters to the original parameters $R_g$ and $R_i$:

4. $\beta$ = the "bifurcation angle," the angle between the pair of branch axes emanating from a bifurcation. It is measured in the "bifurcation plane," which is the plane containing the axis of the initial branch, the bifurcation point, and the two axes of the new branches arising from the bifurcation.

5. $\tau$ = the "angle of twist" between branch axes. This parameter is new and is necessary to model three-dimensional branching, where all the bifurcations no longer occur in a single plane. Consider an initial branch that bifurcates into two new branches and the "bifurcation plane" containing the resulting three branch axes and bifurcation point. Now consider one of the two new branches which itself bifurcates following a period of additional growth. This new bifurcation will also produce a bifurcation plane, but this second bifurcation plane may not lie in the same plane as the initial bifurcation plane. The second bifurcation may occur in a plane oriented at an angle to the original two-dimensional bifurcation plane, resulting in a three-dimensional arrangement of branches. The parameter $\tau$ is the angle between the two bifurcation planes and is itself measured in any plane perpendicular to both the bifurcation planes being considered. Obviously, if all bifurcations in a branching system occur in the same plane (two-dimensional branching geometry), the parameter $\tau$ is equal to zero.

A four-dimensional theoretical morphospace of branching form may be produced by using the parameters $R_g$, $R_l$, $\beta$, and $\tau$ as the dimensional axes. Cheetham and Hayek (1983) produced a series of computer simulations of hypothetical arborescent bryozoan colonies by varying the values of these four parameters. For example, figure 3.3 gives three computer simulations of hypothetical arborescent bryozoans produced by varying the parameter values of $\beta$ and $\tau$ (branch bifurcation angle and twist angle between branches) while holding the values of $R_g$ and $R_l$ constant. Increasing the value of $\beta$ by 30° while holding the value of $\tau$ constant expands the "canopy" area of the "tree" in a side view, as the branches diverge more rapidly away from the central "trunk" (z-axis) of the "tree" (figure 3.3, A versus B). Reducing the value of $\tau$ by 30° while holding the value of $\beta$ constant compresses "canopy" area of the "tree" in the top view, as the branches do not diverge as rapidly away from one another into the y-dimension (figure 3.3, B versus C).

**FIGURE 3.3.** Three simulations of hypothetical arborescent colony form in erect branching bryozoans, taken from the $\beta$-$\tau$-$R_g$-$R_l$ theoretical morphospace of Cheetham and Hayek (1983). Top views of the colony form are given on the left and side views of colony form on the right, where the $x$, $y$, and $z$ axes are relative to the median plane of the colony stem. For simulation A, $\beta = 50°$ and $\tau = 50°$; for B, $\beta = 80°$ and $\tau = 50°$; and for C, $\beta = 80°$ and $\tau = 20°$. $R_g = 1.4$ and $R_l = 1.0$ for all simulations. *Artwork courtesy of A. H. Cheetham. Reprinted from* Paleobiology *and used with permission.*

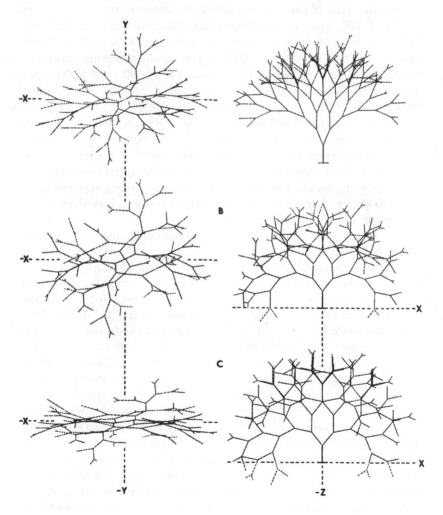

Successive simulations show that colony forms with a maximum radius around the central axis are produced with values of β at 90° and values of τ at 50°. However, colony height is an almost linear function of these same parameters; that is, the higher the values of β and τ are, the lower the height of the colony will be. Cheetham and Hayek (1983) therefore grouped successive values of these two parameters into three classes in terms of their overall effect on colony form. Low-angle values (β < 60°, τ < 30°) produce colony forms of maximum height but also have a major radius of less than 80% and a "three-dimensionality" of less than 60% of the potential colony maximum. At the opposite extreme, high-angle values (β > 90°, τ > 60°) produce colony forms of minimum height, but with a "three-dimensionality" maximum. Intermediate-angle values (β = 60° to 90°, τ = 30° to 60°) produce colony forms with a height range of 85% to 90% of the potential maximum, a major radius that is 80% to 90% of the maximum, and a "three-dimensionality" that is 60% to 90% of the maximum. The intermediate-angle class is also the one with maximum spacing of the growing tips in the colony. Growing-tip spacing declines to less than 80% of maximum in the low-angle class and less than 50% of maximum in the high-angle class.

Cheetham and Hayek (1983) then examined the morphology of existent bryozoan colonies, not only to compare them with the computer simulations of hypothetical colony form, but also to examine the distribution of these real bryozoans in their four-dimensional theoretical morphospace of arborescent bryozoan colony form. Figure 3.4 shows the actual distribution of 17 species of living and fossil adeoniform cheilostome bryozoans in the β–τ plane of the $\beta$–$\tau$–$R_g$–$R_l$ theoretical morphospace. The dashed lines in the morphospace and their intercepts along the two axes delimit the regions of the low-angle, intermediate-angle, and high-angle classes for these two parameters discussed previously. Note that the distribution of actual bryozoan colonies lies almost entirely in the intermediate-angle range for β and in the intermediate-angle to the higher end of the low-angle range for τ (figure 3.4). No bryozoan colonies occur in either the low-β–low-τ or the high-β–high-τ regions of the theoretical morphospace. The region occupied by actual bryozoans in the morphospace is one where colony forms are "minimizing long-term branch interference, while keeping colony height, branch spread, and 3-dimensionality near favorable levels" (Cheetham and Hayek 1983:253).

Cheetham and Hayek (1983) concluded that real arborescent bryozoan colonies have geometries that minimize branch interference in late growth stages. Computer simulations of the range of parameters seen in actual bryozoans show that interbranch distances progressively decrease as branches increasingly converge with growth; thus an emphasis on minimizing branch interference essentially sets an upper limit on growth. Or in other words, "to gain this long-term benefit [of minimum branch interference in late growth] requires adhering to a regular pattern throughout growth" (Cheetham and Hayek 1983:240). That conclusion echoes the earlier conclusion of Cheetham, Hayek, and Thomsen (1980) that a high degree of direct internal control on the branching process appears to occur in arborescent bryozoans.

FIGURE 3.4. The actual distribution of 17 species of fossil and living adeoniform cheilostome bryozoans in the β-τ plane of the four-dimensional theoretical morphospace of Cheetham and Hayek (1983) given in figure 3.3. Geologic ages of the species are indicated by the symbols given in the lower left of the figure. Dashed lines in the morphospace delimit high, intermediate, and low parameter-value regions of the morphospace (see the text for discussion). *Artwork courtesy of A. H. Cheetham. Reprinted from* Paleobiology *and used with permission.*

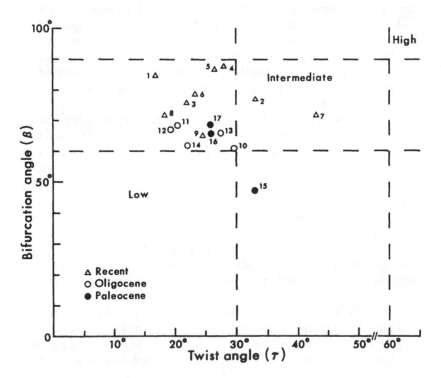

In an interesting aside, Cheetham and Hayek (1983:254) note that arborescent branching in bryozoans "permits colony surface area to expand at an exponential rate," with the obvious benefit of producing the maximum number of zooids for feeding and reproduction but that "erect cheilostome growth is not the only way in which a cheilostome can maximize surface area." Nonarborescent cheilostomes can produce massive hemispherical colonies with surface areas that rival those seen in arborescent colonies. The advantage of arborescent growth is that the adeoniform colonies achieve their large surface areas with only 2% to 3% of the internal colony volume required by the hemispherical colonies.

## THEORETICAL MORPHOSPACES OF BRANCHING FORM IN TERRESTRIAL PLANTS

A series of early studies of branching form in plants do not explicitly construct theoretical morphospaces, but they do follow the "theoretical morphospace methodology" in that they create a geometric continuum of hypothetical plant-branching morphologies, morphologies that are mathematical constructs with no real biological existence, and then they compare actual plant branches with the spectrum of form seen in the hypothetical branching forms. They also complete the final important step in theoretical morphospace methodology (see chapter 2) in that they analyze the functional significance of the match seen between real plant branches and one of their many hypothetical branching forms.

Leopold (1971) analyzed branching patterns in trees, using the "ordinal" classification of branches originated by Horton (1945) and modified by Strahler (1952) for river systems (where branches are classified as "first order," "second order," and so on. We already have discussed this branch classification system, as it was also used by Cheetham, Hayek, and Thomsen [1980–1983] in their series of studies of bryozoan morphology). Leopold found that similar to branching in river systems, a logarithmic relationship exists between branch order and branch length and numbers in the trees he studied. He concluded that the observed pattern of branching in trees exhibits a "tendency for efficiency by the minimization of the total length of all the branches" (Leopold 1971:351).

Leopold's study (1971) is actually a combination of morphometrics and functional morphology. I include it here with theoretical morphology because of his discussion of potential types of branching geometry in plants and his simulation of branching properties by a process of random walk (Leopold 1971:348–351). In these simulations he created a series of hypothetical two-dimensional "tinker-toy trees" (his term) and then explored the relationship between their branch orders and branch lengths and numbers. He found that the branches of his random-walk trees converged on the same logarithmic relationship as seen in real trees, concluding that "this logarithmic relationship is one of optimum probability" (Leopold 1971:353).

A study of plant form that is more explicitly theoretical morphologic in approach is that of Honda (1971), who stated at the onset that his study was an attempt to model "the multifarious form of erect trees by a few parameters." Honda's computer simulations of three-dimensional branching form in hypothetical trees are elegant and antedate by a dozen years those developed for arborescent animals. The model begins with a primary structural element, termed the "mother branch," which branches in a dichotomy to produce two new structural elements, termed "daughter branches." The branching process is not a simple bifurcation in which one daughter branch diverges at an angle to the left of the line of axis of the mother branch and the other daughter branch diverges to the right of the line of axis of the mother branch by the same angle (see the earlier discussion in this chapter of dichotomous branching in graptoloids). Instead, in Honda's model, the angle of divergence from the line of axis of the mother branch is $\theta_1$ in one daughter branch and $\theta_2$ in the other daughter branch (in a simple bifurcation, $\theta_1$ is equal to $\theta_2$). The lengths of the two daughter branches are calculated as ratios, $R_1$ and $R_2$, respectively, of the length of the mother branch, which always is longer. To simulate an erect "trunk" growing perpendicular to the ground (and parallel to the line of action of gravity), $\theta_1$ is set equal to zero. That is, one daughter branch diverges by $\theta_2$ from the line of axis of the mother branch, whereas the other "daughter branch" in the dichotomy does not diverge at all but continues to grow in the line of axis of the mother branch (thus constituting the growing trunk of the tree). To move the simulation into the third dimension, a divergence angle, termed $\alpha$, is introduced to measure the placement of branches arising from the central trunk. A single branch arising from the trunk defines a plane that contains the line of axis of the trunk and of the branch. A second branch higher up the trunk defines another plane, which may not be the same plane as that containing the first branch. The angle between these two branch planes, taken in a plane perpendicular to the trunk, is the divergence angle.

Although he used the theoretical morphospace methodology, Honda (1971) did not explicitly create a theoretical morphospace of hypothetical tree form. Such a theoretical morphospace could easily be constructed, however. It would have five dimensions, defined by the parameters $\theta_1$, $\theta_2$, $R_1$, $R_2$, and $\alpha$. Honda's paper has six photographic plates, containing 24 elegant computer simulations of hypothetical tree form, produced by varying the values of those five parameters. (A brief aside: unfortunately, those computer simulations are all white on black rather than the black-on-white format seen in figure 3.3 for bryozoans. To try to include them here would necessitate including photographic plates in the book, which are expensive. Instead, I include in figure 3.5 some monopodial treelike simulations that are similar to those produced by Honda's method but that were produced by the totally different simulation technique of Lindenmayer systems, or "L-systems" [Prusinkiewicz and Lindenmayer 1990, 1996]. We shall encounter L-systems again in chapter 10, when we consider theoretical morphogenetic models of branching systems.) It would be simple indeed to construct a series of two-dimensional parameter-pair "slices" through the five-dimensional $\theta_1$–$\theta_2$–$R_1$–$R_2$–$\alpha$ morphospace, using Honda's simulations to illustrate the hypothetical morphologies found at selected regions in the morphospace (similar to the format seen in figures 3.1 and 3.2 of this chapter).

Honda (1971) does complete the theoretical morphospace methodology by including a very brief discussion both of real plants with respect to the hypothetical plants created by computer simulation and of the potential functional significance of variation in tree form. For example, degree of "axiality" refers to how clearly a plant develops a central axis or trunk, and Honda ranked the following real plants in terms of decreasing axiality: fir, larch, ginkgo, birch, beech, maple, and azalea. Honda notes that the observed variation in the degree of axiality in real plants can be simulated by the parameters $\theta_1$ and $\theta_2$. Parameters $R_1$ and $R_2$ determine the degree of apical dominance, and variation in their values produces trees ranging from conical to flat crowned.

Honda created an explicit and very interesting theoretical morphospace of branching form in a later paper (Honda and Fisher 1978). The two branch angles $\theta_1$ and $\theta_2$ are used to define the two dimensions of the theoretical branching space, and a third dimension graphs a geometric property of the two-dimensional theoretical morphospace itself. This geometric property is the "relative leaf area" produced in a hypothetical tree by varying the parameter values of $\theta_1$ and $\theta_2$ (figure 3.6). The geometry of the hypothetical tree itself is important in determining relative leaf area as well. Honda and Fisher chose to simulate tree forms similar to those found in the tropical tree *Terminalia catappa*. Trees of this type have an erect central axis, or trunk, with tiers of three to five lateral branches originating from the trunk in a periodic fashion. The resulting tree form has a "pagoda-shaped" crown, and each of the tiers in the crown can be treated as a monolayer of leaves, as most of the leaves occur in a flattened layer at the distal ends of the branches.

FIGURE 3.5. Computer-produced hypothetical trees. These monopodial treelike simulations are similar to those produced by Honda 's method (1971) but were produced by the totally different simulation technique of Lindenmayer systems, or "L-systems." *From Prusinkiewicz and Linden-mayer 1990, 1996. Copyright © 1990 by Springer-Verlag and Przemyslaw Prusinkiewicz and used with permission.*

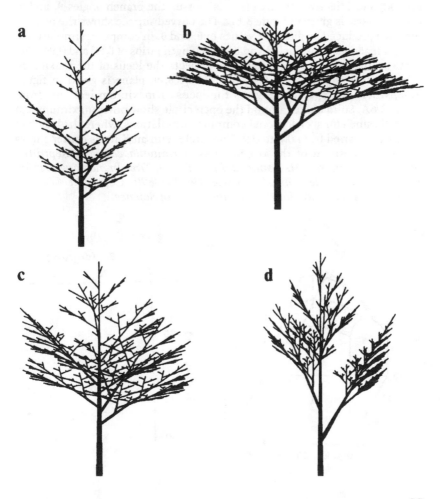

Given such a "pagoda-shaped" tree geometry, the relative leaf area in the crown can be calculated as a function of the five variables: $\theta_1$, $\theta_2$, $R_1$, $R_2$, $\alpha$ and the size of the "leaf disk radius" (a measure of the leaf's surface area). Honda and Fisher (1978) chose a leaf disk radius of 0.8, and branch length ratios ($R_1$ and $R_2$) of 0.87 and 0.94, on empirical grounds, as these values best matched those seen in *Terminalia* for these three variables. The other two variables ($\theta_1$ and $\theta_2$) were allowed to vary in successive computer simulations, in each of which the resultant relative leaf area was calculated. The result is the surface given in figure 3.6A. The maximum leaf area is attained when $\theta_1$ and $\theta_2$ are 24.4° and −41.4°, respectively.

**FIGURE 3.6.** Effective leaf area ($z$-axis) versus the branch angles $\theta_1$ and $\theta_2$ ($x$-and $y$-axes) is given in figure 3.6A. The curved surface shows the relative leaf area produced by different values of $\theta_1$ and $\theta_2$ in computer simulations, using a leaf disk radius of 0.8 and branch length ratios of 0.87 and 0.94. The vertical line from the curved surface projects the locus of the maximum effective leaf area into the $\theta_1$-$\theta_2$ plane. The $\theta_1$-$\theta_2$ plane is given in figure 3.6B, where the open circle gives the locus of maximum leaf area from figure 3.6A. Solid circles around the open circle show other maximum leaf areas obtained for six additional computer simulations, where the leaf disk radius was varied from 0.7 to 0.9. The circle with an X in its center shows the actual position of the tropical tree *Terminalia catappa. Reprinted with permission from H. Honda and J. B. Fisher, "Tree branch angle: maximizing effective leaf area," Science 199:888–890. Copyright © 1978, American Association for the Advancement of Science.*

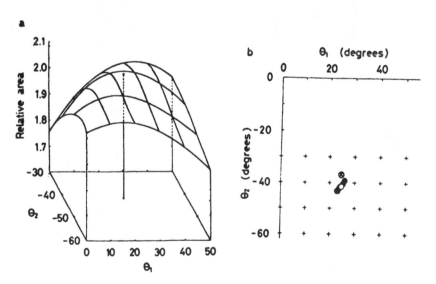

The vertical dimension in Honda and Fisher's (1978) theoretical morphospace is unusual in that it is not a separate morphological dimension but is, rather, a geometric measure of properties of the $\theta_1$-$\theta_2$ two-dimensional theoretical morphospace. If one equates "relative leaf area" with "fitness," then figure 3.6A meets the criteria of an adaptive landscape, listed in chapter 2 (cf. figures 2.1 and 3.6A). The "adaptive peak" in the theoretical morphospace occurs at the point where $\theta_1$ and $\theta_2$ are 24.4° and −41.4°, where the maximum leaf area is attained and hence the maximum photosynthetic potential for the tree. The values of $\theta_1$ and $\theta_2$ found in the actual tree *Terminalia catappa* are very close to those of the "adaptive peak," as shown in figure 3.6B. A selectionist interpretation of that correspondence would be, of course, that natural selection has indeed favored a tree geometry producing the maximum photosynthetic potential. The geometry of the relative leaf area surface (figure 3.6A) appears to be relatively robust, in that the maximum leaf area position does not shift much in the $\theta_1$-$\theta_2$ theoretical morphospace if different values of leaf disk radius are used in the computer simulations (figure 3.6B).

A much more generalized, and hence more inclusive, theoretical morphospace of erect branching form in plants was created by Niklas and Kerchner (1984). The parameters of the morphospace are as follows:

1. $p$ = the "probability of branching termination" and is a measure of the frequency of bifurcation or branching. The value of $p$ ranges from 0.0, the highest frequency of bifurcation, to 0.9, the lowest frequency of bifurcation.
2. $\phi$ = the "branching angle" or "bifurcation angle." The parameter $\phi$ is the same as the parameter $\theta$ in Honda (1971) and Honda and Fisher (1978). The terminology of its usage is a little different, however. This condition, where $\phi_1$ is equal to $\phi_2$, is termed "isobifurcating" by Niklas and Kerchner (1984). The dichotomous branching condition, where $\phi_1$ is not equal to $\phi_2$, is termed "unequal bifurcating." "Unequal bifurcation angles" result when there is a "main tree" axis and "lateral branch" axes and their branching angles are denoted $\phi_1$ and $\phi_2$, respectively. The condition where $\phi_1$ is equal to zero is termed "pseudomonopodial" and is used in the computer simulations to produce a central trunk axis, as in Honda (1971).
3. $\gamma$ = the "rotation angle" of the branch. The parameter $\gamma$ specifies the angle of rotation of the "daughter branches" about

57

the line of axis of the "mother branch" and is similar to the "angle of twist" parameter $\tau$ of Cheetham and Hayek (1983).

Niklas and Kerchner's (1984) theoretical morphospace has only three dimensions in that it was designed to examine hypothetical plant form in isobifurcating branching systems ($\phi_1$ equal to $\phi_2$) and pseudomonopodial treelike systems (where $\phi_1$ is equal to zero), thus allowing the two branch-angle dimensions ($\phi_1$ and $\phi_2$) to be collapsed into one (figure 3.7). For purposes of simplicity in the computer simulations, branch lengths were kept constant and numbered in an ordinal system (first-order branches distalmost, uniting to form second-order branches, and so on), where the maximum number of sequential branching events was set at 10. Thus the actual size of the hypothetical plant produced by simulation is a function of $p$, where simulations with a high probability of terminating bifurcation remain small.

Niklas and Kerchner (1984) then explored some of the geometric properties of their theoretical morphospace. They first divided each dimensional axis of the three-dimensional theoretical morphospace, or "cube," into 10 equal increments (figure 3.7). Thus on any given face of the cube there are 100 square cells, and throughout the total volume of cube there are 1,000 cubic cells, which Niklas and Kerchner termed "coordinate sets." For each "coordinate set," 10 branching simulations were produced (that is 10,000 simulations).

For each branching simulation, two metrics were computed: the "photosynthetic efficiency" of the morphology (the metric $I$) and the "total moment arm" of the morphology (the metric $M$). The efficiency of photosynthesis, $I$, is a function of the surface area of the photosynthesizing organ relative to its volume. If the photosynthesizing organ is the branch surface itself, it can be shown that cylindrical branches are more photosynthetically efficient than other geometries (Niklas and Kerchner 1984). The photosynthetic efficiency of placement of the cylindrical branch surfaces in an erect, treelike branching system is a function of the branching angle, $\phi$, and the rotation angle, $\gamma$, of the branching system, as well as the number of branches produced in growth, which is a function of $p$. Erect treelike growth also has its mechanical, as well as area-volume, constraints (McMahon and Kronauer 1976). The central axis, or trunk, of a tree experiences both compressive and tensile stresses produced by the bending moment generated by the weight of the branches and axis of the tree. The stress related to the total moment arm, $M$, is likewise a function of $\phi$, $\gamma$, and $p$.

**FIGURE 3.7.** Niklas and Kerchner's (1984) $p$-$\gamma$-$\phi$ theoretical morphospace of erect branching form in plants. The parameters of the morphospace are $p$, the probability of branching termination (a measure of the frequency of bifurcation); $\phi$, the branching angle; and $\gamma$, the rotation angle of the branch. Each axis is divided into 10 equal measurement increments, producing 1,000 cubic cells or "coordinate sets" of parameter combinations in the morphospace. For each coordinate set, 10 branching simulations were produced by computer, and from those simulations, the average photosynthetic efficiency ($I$), total moment arm ($M$), and the ratio $I/M$ for that coordinate set were determined. *From Niklas and Kerchner 1984. Reprinted from* Paleobiology *and used with permission.*

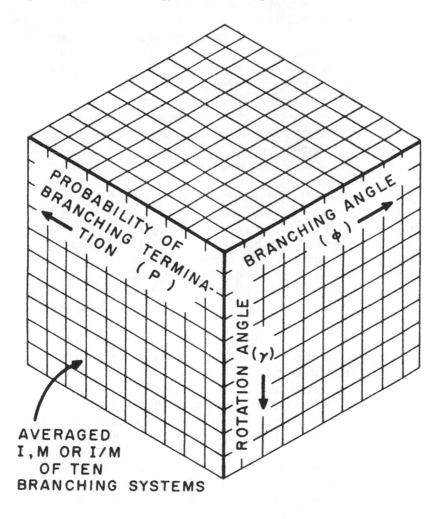

The two metrics were also combined as a ratio ($I/M$) to express the "branching efficiency" of the simulated morphology. Low values of the ratio $I/M$ indicate "inefficient branching patterns" having low photosynthetic efficiencies and high bending stresses due to large moment arms. High values of $I/M$ indicate "efficient branching patterns." Last, the values of $I$, $M$, and $I/M$ obtained from each of the 10 simulations run for each coordinate set were averaged to obtain a single mean value of these metrics for each of the 1,000 coordinate sets in the theoretical morphospace.

The spatial distribution of values of the ratio of photosynthetic efficiency to total moment arm in the $p$-$\gamma$-$\phi$ theoretical morphospace for pseudomonopodial (treelike) branching forms is shown in figure 3.8A and B, and the distribution of values of the photosynthetic efficiency metric alone is shown in figure 3.8C and D. (A brief aside: note that the directions of increase in the morphospace's parameters in the view given in figure 3.8 are different from those given in the generalized summary cube in figure 3.7. The direction of $\phi$ is the same, but the increase directions of $p$ and $\gamma$ are reversed.) Selected computer simulations of branching geometries are also given for various regions in the theoretical morphospace. Values of $I/M$ and $I$ given in figure 3.8 are scaled in percentages of their maxima, $I/M_{max}$ and $I_{max}$. Solid regions in the morphospace in figure 3.8A show the distribution of branching geometries with $I/M$ scores of 80%. In figure 3.8B, the right rear region of the morphospace contains geometries with $I/M$ scores of 90% or higher, whereas the "suspended" region (upper left of the morphospace) has geometries with the lowest scores (60%) in the morphospace. In general, the ratio of $I$ to $M$ increases as $p$ and $\phi_2$ increase and as $\gamma$ decreases. Thus the region of the theoretical morphospace having treelike geometries with the highest branching efficiency lies in the lower rear corner of the cube, extending from the rearmost corner forward to the lower right side corner and upward to the upper right side corner (figure 3.8B). With decreasing values of branching efficiency, this region expands upward toward the top of the cube and forward toward the front corner of the cube (figure 3.8A).

The spatial distribution of regions of higher photosynthetic efficiency in the theoretical morphospace is different from that seen for overall branching efficiency. The solid region in figure 3.8C contains branching geometries with $I$ scores of 71% through 89%, and the open region in the foreground has geometries with $I$ of 70% or lower. In figure 3.8D, the left rear region of the morphospace contains geometries with $I$ scores of 90% or higher. Thus just about the entire morphospace is characterized by branching geometries with photosyn-

FIGURE 3.8. The distribution of values of the ratio of photosynthetic efficiency to total moment arm (A and B) and of photosynthetic efficiency alone (C and D) in the $p$-$\gamma$-$\phi$ theoretical morphospace for pseudomonopodial (treelike) branching forms. Values of $I/M$ and $I$ scaled in percentages of their maxima, $I/M_{max}$ and $I_{max}$. Solid regions in the morphospace in figure 3.8A show the distribution of branching geometries with $I/M$ scores of 80%. In figure 3.8B, the right rear region of the morphospace contains geometries with $I/M$ scores of 90% or higher, whereas the "suspended" region (upper left of the morphospace) has geometries with the lowest scores (60%) in the morphospace. The solid region in figure 3.8C contains branching geometries with $I$ scores of 71% through 89%, and the open region in the foreground has geometries with $I$ of 70% or lower. In figure 3.8D, the left rear region of the morphospace contains geometries with $I$ scores of 90% or higher. Selected computer simulations of branching geometries in the morphospace are given for each figure. *From Niklas and Kerchner 1984. Reprinted from* Paleobiology *and used with permission.*

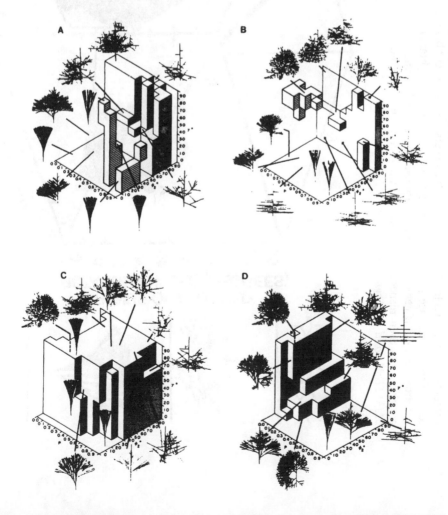

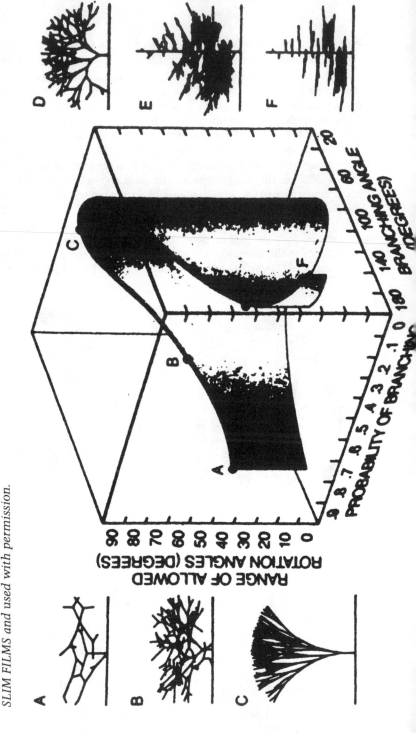

FIGURE 3.9. A computer-generated evolutionary trajectory of increasing "fitness" (in terms of maximizing $I$ and minimizing $M$) in Niklas and Kerchner's $p$-$\gamma$-$\phi$ theoretical morphospace (1984). Starting with plant simulation at position A, the computer was programmed to search for the nearest coordinate sets with a "fitter" branching geometry (greater $I$ or lesser $M$, or both). Moving to the fitter coordinate set, the search process for an even fitter branching geometry is repeated, resulting in the trajectory to the fittest branching geometry in the morphospace. Selected computer simulations of branching geometries along the trajectory are given, and their respective positions in the morphospace are indicated by letters. *Modified from Niklas 1986. Artwork by SLIM FILMS and used with permission.*

thetic efficiencies of 71% or higher. Only the small region in the front corner of the cube, extending over to the left side corner, is characterized by branching geometries with lower photosynthetic efficiencies. Note in conclusion, however, that the region of the theoretical morphospace having branching forms with the highest photosynthetic efficiency (figure 3.8D) is not the same as the region having treelike geometries with the highest branching efficiency (figure 3.8B).

Niklas and Kerchner (1984) did not plot measurement data for the positions of actual trees in their theoretical morphospace. In the spirit of Raup (1966), they did outline general regions in the morphospace that they considered to correspond to specific plant groups, based on the morphology of the computer simulations alone (an aside: see figure 4.3 for the distribution of molluscan groups in Raup's $W–D–T$ theoretical morphospace, regions that were drawn without reference to actual measurement data). Three generalized botanical regions are recognized in the theoretical morphospace: a gymnosperm or *Picea*-like branching region, a pteridophyte or *Psilotum*-like branching region, and an angiosperm or *Quercus*-like branching region.

The gymnosperm region in the theoretical morphospace corresponds to the 90% $I/M$ ratio region illustrated in figure 3.8B and the expansion of this same region into the 80% $I/M$ ratio region in figure 3.8A. This region in general is characterized by high values of $\phi_2$, producing treelike morphologies with the lateral branches originating at high angles (if not right angles) to a well-developed central trunk axis, such as is seen in many modern pines and other conifers. The pteridophyte region corresponds to the 80% $I/M$ ration region located in front of the gymnosperm region in figure 3.8A, that is, the region with low values of $\phi_2$ (10° to 40°) and high values of $p$ (0.6 to 0.9). Computer simulations in this region produce erect conical-brush–branching morphologies, such as is seen in some living and fossil members of the ancient pteridophyte botanical group. Both the pteridophyte and gymnosperm regions are regions of high branching efficiency (figure 3.8A and B). The pteridophyte region has the lowest photosynthetic efficiency, however, and the gymnosperm region has only modest photosynthetic efficiency (cf. figure 3.8A, B, C, and D).

In contrast, the angiosperm region in the theoretical morphospace corresponds to the 90% $I$ value region shown in figure 3.8D. This region is characterized by low values of $p$ (0.0 to 0.3), and it expands in volumetric extent in the theoretical morphospace with decreasing values of $\gamma$. Computer simulations in this region produce bushy arborescent morphologies similar to many flowering plants (figure 3.8D). While having the maximum photosynthetic efficiency, such morphologies also have very high total moment arms, hence low branching efficiency (cf. figure 3.8D and B).

A trade-off thus appears to exist between maximizing photosynthetic efficiency, on the one hand, and minimizing bending stress due to total moment arm length, on the other. Based on their theoretical morphologic analysis of branching form, Niklas and Kerchner (1984) proposed a potential evolutionary series of geometric changes that would enhance photosynthetic efficiency while minimizing total moment arm in plants. These potential evolutionary changes are termed "optimizing geometric adjustments" (Niklas and Kerchner 1984:98) and are (1) a transition from diffuse growth to apical growth; (2) repeated bifurcation of cylindrical axes (setting $\phi_1$ equal to $\phi_2$ for an initial isobifurcating system); (3) vertical orientation of axes (and adding $\gamma$ for three-dimensional branch deployment); (4) adjustment of bifurcation and rotation angles toward the path of the sun (adjusting $\phi$ and $\gamma$); (5) adjustment of the frequency of bifurcation, resulting in the production of an "overtopped" branching pattern (making $p$ a function of growth stage); (6) a tendency to produce a "main" vertical axis, bearing "lateral branching" systems that acropetally decrease in size (decreasing $\phi_1$ eventually to $0°$ and increasing $p$ in lateral branches); (7) a tendency to "planate" lateral branching systems (reducing $\gamma$ eventually to $0°$ in lateral branches); and (8) the support of lateral, planated branching systems at right angles to the main vertical axis (increasing $\phi_2$ eventually to $90°$ in lateral branches).

The hypothetical trajectory of the evolution of plants through the $p$-$\gamma$-$\phi$ theoretical morphospace through the process of natural selection for Niklas and Kerchner's optimizing geometric adjustments (1984) is summarized in figure 3.9. Note that view of the theoretical morphospace given in figure 3.9 has been rotated from that in figure 3.8 in order to give a better perspective on the shape of the evolutionary pathway through the morphospace. The vertical axis ($\gamma$) has the same orientation as in figure 3.8. The cube given in figure 3.9 has been rotated $180°$ about the vertical axis from its orientation in figure 3.8, so that the rearmost edge now is in the front, and vice versa. That is, the lower corner of the cube where $p$ is equal to zero and $\phi$ is maximum is in the front in figure 3.9 but in the back in figure 3.8. And last, note that the maximum value of the branching angle ($\phi$) has been allowed to increase to $180°$ in figure 3.9, rather than stopping at $90°$ as in figure 3.8.

The trajectory in the theoretical morphospace shown in figure 3.9 was actually produced by a computer simulation of evolution via the process of natural selection (Niklas 1986; see also Swan 1990a). Maximum "fitness" was defined in terms of maximum photosynthetic efficiency ($I$) and minimum bending stress due to total moment arm ($M$). Starting with plant simulation at position A, the computer was programmed to search for the nearest coordinate sets with a "fitter" branching geometry (greater $I$ or lesser $M$, or both). Moving to the fitter coordinate set, the search for an even fitter branching geometry was repeated, resulting in the trajectory to the fittest branching geometry in the morphospace.

Selected computer simulations of branching geometries along the evolutionary trajectory are given in figure 3.9 and their respective positions in the morphospace indicated by letters. Note that the general evolutionary trend follows that of the eight optimizing geometric adjustments discussed earlier. The most primitive plant (position A) is low and sparsely branched. It evolves into plants that have more branches (position B) and that are more vertical (position C). The branches of the plant begin to fill three-dimensional space more efficiently, maximizing exposed surface area (for photosynthesis) while minimizing the shading of the branches of one another (position D). The emergence of a well-developed central trunk axis appears in position E, from which the lateral branches originate at right angles in position F. And last, the lateral branches in position F have become planate.

# IV

# Spirals and Shells I: Theoretical Morphospaces of Univalved Accretionary Growth Systems

In this and the next chapter we shall examine theoretical morphospaces created for organisms that use accretionary growth systems. Nature's beautiful spiral morphologies are often the result of accretionary growth systems, and many elegant mathematical models have been created for spiral morphogenesis. My goal in this chapter is to introduce some of the aspects of each theoretical morphospace in as general terms as possible and not to lose the reader in a tangle of mathematical and parameter specifics; thus the in-depth analysis of the mathematical models used to produce these morphospaces will be deferred to chapters 8 and 9. But if the reader is intrigued by any of the morphospace aspects here discussed and wishes to jump directly into the mathematics of the model under discussion, please feel free to turn to the relevant section in chapters 8 and 9.

In our day-to-day experience, we perceive the universe around us in terms of the dimensions of length, width, height, and time or, in more abstract terms, in terms of the four dimensions $x$, $y$, $z$, and $t$. Therefore, in the following discussions, I shall attempt to translate the various authors' original formulations of their theoretical morphologic models in terms of these four dimensions.

# THEORETICAL MORPHOSPACES OF
# UNIVALVED SHELL FORM

David M. Raup created the first computer-generated theoretical morphospace for univalves—as outlined in the papers by Raup and Michelson (1965) and Raup (1966)—a morphospace that followed from Raup's earlier computer simulation studies of shell form generation (1961, 1962). The mathematical model used to produce Raup's theoretical morphospace falls in the general class of "generating-curve models with fixed-reference frames" (see chapter 9). We begin with a geometric figure (most often a circle or an ellipse) termed a "generating curve." The term "generating curve" is somewhat confusing, as most people think of a curve as a segment of a geometric figure and not an entire figure such as a circle. This term is firmly entrenched in the literature of theoretical morphology, however, and I will not attempt to replace it here with alternatives such as "aperture" or "tube cross section."

The initial generating curve is located in the x-z plane, with the center of the generating curve located on the x-axis (in the case of a circle, the x-axis bisects it, as in the circle on the right in figure 4.1). The plane containing the generating curve is then rotated around the z-axis, which is termed the "coiling axis." Two changes take place during this rotation: (1) the generating curve becomes larger (in the case of a circle, its radius increases), and (2) the generating curve moves in relation to the z-axis. In simple planispiral growth, the generating curve moves away from the z-axis, and its center leaves the trace of a two-dimensional spiral in the x-y plane. In somewhat more complex helicospiral growth, the generating curve moves both away from the z-axis and along the z-axis, and its center produces the trace of a three-dimensional spiral in x-y-z space. In the case of a circular generating curve, the initial two-dimensional generating curve produces a three-dimensional "coiled cone": a "cone" due to the radial expansion of the circle in time, and "coiled" because this cone is wrapped around the z-axis with rotation.

In Raup's model, the two changes that take place in the generating curve with time are always measured with reference to the z-axis, or coiling axis; hence this type of model is a "fixed-reference frame" model (see chapter 9). In planispiral growth, two vectors can be used to parameterize the changes that take place in the generating curve as it is rotated about the z-axis: a vector taken in the x-y plane from the z-axis to the outer margin of the generating curve (the margin of the generating curve farthest from the $z$-axis), which I will term here $v_O$ for "outer vector," and a vector taken in the $x$-$y$ plane from the $z$-axis to the inner margin of the generating curve (the margin of the generating curve closest to the $z$-axis), which I will term here $v_I$ for "inner vector" (see the circular generating curve on the right in figure 4.1). In

FIGURE 4.1. A vector representation of Raup's (1966) parameters in $x$-$y$-$z$ space (the $y$-dimension is perpendicular to the page). On the right, a circular generating curve is centered on the $x$-axis. The inner margin of the generating curve is determined by the vector $v_I$, and the outer margin by the vector $v_O$, both of which are normal to the $z$-axis. On the left, the generating curve has been rotated 180° about the $z$-axis, has expanded in size, and has translated down the $z$-axis. The vectors determining the inner and outer margins of the generating curve are now located at an angle to the $z$-axis. The vectors $v_I$ and $v_O$ are now taken as projections of the marginal vectors on the $x$-axis and a new vector, $v_T$, is taken as the projection of the center of the generating curve onto the $z$-axis. In the generating curve in both positions are two additional vectors not found in Raup's parameters: $v_R$, from the center of the generating curve to its margin, and $v_D$, from the center of the generating curve into the $y$-dimension (perpendicular to the page). See the text for discussion.

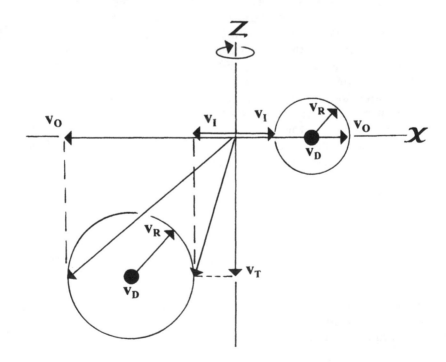

simple planispiral growth, the vectors $v_O$ and $v_I$ remain in the $x$-$y$ plane. In more complex helicospiral growth, these two vectors lift out of the $x$-$y$ plane and thus have $x$-$y$-$z$ coordinates. In helicospiral growth, we can now still define $v_O$ and $v_I$ in terms of the $x$-$y$ coordinates of these three-dimensional vectors (i.e., the projection of the three-dimensional vectors into the two-dimensional $x$-$y$ plane) while taking the $z$-coordinate to define a new vector, here termed $v_T$ for "translation vector" to measure the translation of the generating curve along the $z$-axis (see the circular generating curve on the left in figure 4.1).

Raup defined four geometric parameters from his model, which he then used to define the four dimensions of his morphospace:

1.  $W$ = the "whorl expansion rate" (of the generating curve) from the coiling axis. The whorl expansion rate measures the rate of the movement of the generating curve away from the $z$-axis with time. In terms of the vectors defined earlier, the parameter $W$ measures the rate of magnitude change in $v_O$ relative to its change in direction with time (i.e., its rate of rotation around the $z$-axis).

2.  $D$ = the "distance from the coiling axis to the generating curve." The parameter $D$ measures the increase in size of the generating curve with time by specifying the ratio between changes in the distance of offset of the generating curve from the $z$-axis (changes in magnitude of $v_I$) relative to the changes in the whorl expansion rate (changes in magnitude of $v_O$).

3.  $T$ = the "translation rate" (of the generating curve) along the coiling axis in helicospiral growth. The translation rate measures the rate of movement of the generating curve along the $z$-axis with time. In vector terms, the parameter $T$ measures the rate of magnitude change in $v_T$ relative to the rate of magnitude change in $v_O$.

4.  $S$ = the "shape" of the generating curve. The parameter $S$ is left open for definition by the individual researcher but in general is taken as the ratio between a minor axis and a major axis drawn in the generating curve. For a circle, $S$ is equal to 1, as all diameters drawn in the circle are of the same magnitude.

At this point, I would like simply to mention to the reader that there is a bit of a problem in the mathematical formulation of Raup's parameters $D$ and $T$ and to move on with our consideration of the theoretical model itself. In short, defining the $D$ parameter as the magnitude ratio of the two vectors $v_I$ and $v_O$, rather than a simple measure of offset from the z-axis (magnitude of $v_I$), and scaling the $T$ parameter as the rate of magnitude change in $v_T$ relative to the rate of magnitude change in $v_O$, rather than simply the rate of magnitude change in $v_T$, causes a dimensional problem, which will be discussed in more detail later.

Raup formulated his model in terms of continuous differential equations and specified that the values of the model parameters follow a simple exponential growth function, where the rate of change of the parameters remains constant with time. This specification results in the condition of isometric growth, and the spirals produced by the track of the generating curve through x-y-z space are logarithmic (a result explained in more detail in chapter 8).

The results of systematically varying the values of Raup's three principal parameters $W$, $D$, and $T$ in the computer simulation of univalved shell morphology are given in figure 4.2. The simulated morphologies in figure 4.2 are produced by analog computer (rather than digital) and are displayed on an oscilloscope screen (rather than a laser printer) so have the appearance of radiographs. Perhaps the most striking morphologic result to those who view this figure for the first time is that produced by variation in the parameter $T$. What appears to be a radiograph of a very real ectocochleate cephalopod at $T$ equal to zero is incrementally transformed into an apparent radiograph of a very real orthostrophic gastropod at $T$ equal to 4. 0. And that striking transformation between two entirely different classes of animals in the real world is brought about in theoretical morphologic simulation by variation in a single geometric parameter, $T$.

The morphologic results of variation in the parameters $W$ and $D$ are a bit more subtle but equally profound. In figure 4.2 the ectocochleate cephalopod at $W$ equal to 4.5 is transformed into what appears to be one of the valves of a terebratulide brachiopod at $W$ equal to $10^2$, a morphologic transformation that represents a phylogenetic jump between two entirely different phyla of animals. Finally, our ectocochleate cephalopod appears to have a convolute shell at $D$ equal to zero (like the living nautiloid *Nautilus*), but this is transformed in an advolute shell at $D$ equal to 0.3 (like the extinct ammonoid *Perisphinctes*).

Raup then treated each of the parameters $W$, $D$, and $T$ as independent spatial dimensions by placing them orthogonal to one another to produce a three-dimensional theoretical morphospace of univalved shell form (figure 4.3). Figure 4.3 is arguably the most famous illustration ever produced in theoretical morphology, as it has been reprinted or reproduced, in one form or another, in countless scientific publications since 1966. It also promptly gained the nickname of "the Cube" among many theoretical morphologists, for obvious reasons.

Why has the "Cube" been so significant to the science of morphology, so that it has been cited so many times? To answer that question, let us begin with two morphologic features, evolutionary results, which can be seen in the three-dimensional $W$–$D$–$T$ morphospace in figure 4.3 that is not at all obvious in the separate $W$, $D$, and $T$ parameter variation results given in figure 4.2. Although he had actual measurement data only for the ammonoids at the time, Raup (1966) outlined in the "Cube" his estimate of the expected distribution of typical morphologies seen in the shells of the two univalved molluscan

FIGURE 4.2. Hypothetical univalved shell forms, produced by analog computer, illustrating the morphologic effects of changing the parameter values of the three geometric parameters $W$, the whorl expansion rate from the coiling axis, $D$, the distance from the coiling axis to the generating curve, and $T$, the translation rate along the coiling axis. *From Raup 1966. Used with permission of SEPM—Society for Sedimentary Geology (formerly Society of Economic Paleontology and Mineralogy).*

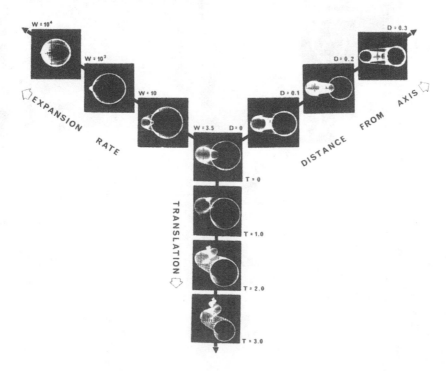

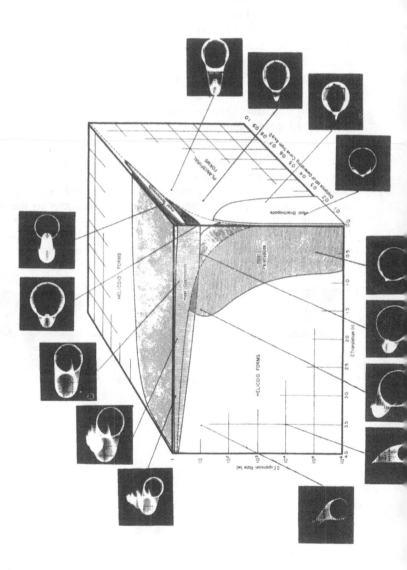

FIGURE 4.3. Raup's famous *W–D–T* univalved theoretical morphospace (1966), also known as the "Cube." Values of the geometric parameter *T* are scaled along the x-axis, values of *D* along the y-axis, and values of *W* along the z-axis (see figure 4.2 for an illustration of the morphological effects of variation in these three parameters). Selected hypothetical shell forms, produced by analog computer, are given around the margin of the figure to show the distribution of morphologies in the theoretical morphospace. Raup's estimate of the distribution of the characteristic morphologies found in gastropods, pelecypods (= bivalve molluscs), ammonoids, and brachiopods is indicated in the figure. *From Raup 1966. Used with permission of SEPM—Society for Sedimentary Geology (formerly Society of Economic Paleontology and Mineralogy).*

groups of gastropods and ammonoids, and the separate valves of the two bivalved groups of brachiopods and bivalve molluscs, based on the observed range of morphologic variation produced in the computer simulations. The first morphologic feature can be clearly seen in figure 4.3: *most of the* W–D–T *theoretical morphospace is empty.* It was expected beforehand that some entirely conceivable morphologies, which can actually be produced in computer simulation by varying the model parameter values of *W, D,* and *T,* might not have actually existed in nature, so that parts of the theoretical morphospace might be empty. The fact that most of the theoretical morphospace is empty was a confirmation of what previously had only been conjectured from the concept of the adaptive landscape, as discussed in chapter 2. (A brief aside: Kauffman [1993:19] considers empty morphospace to be "filled with red herrings" and points out that in a hypothetical morphospace in which no constraints at all exist, a hypothetical evolutionary walk consisting of purely random steps and branchings will still fill only a very small region of a high-dimensional space. Note, however, that the empty regions in figure 4.3 occur in a three-dimensional space, which is a theoretical morphospace of very low dimensionality. Kauffman's main point is that empty morphospace in itself does not constitute evidence for evolutionary constraint. An actual example of evolutionary constraint in the W–D–T theoretical morphospace, demonstrable on purely geometric grounds, will be discussed shortly.)

The second morphologic feature seen in figure 4.3 is that the four taxonomic groups (gastropods, ammonoids, brachiopods, bivalves) are *confined to nonoverlapping regions* in the theoretical morphospace. These two results of the theoretical morphologic analysis led Raup to conclude: "Clearly, the distribution of actual species in not random" (Raup 1966:1185). The evolutionary significance of this conclusion will be explored in greater detail later in this chapter.

Figure 4.3 contains another important morphologic feature, although it is more subtle than the empty areas and nonoverlapping regions in the theoretical morphospace. There are *geometric constraint boundaries* in the theoretical morphospace. The particular geometric constraint boundary in the W–D–T theoretical morphospace is the boundary between the region where "whorl overlap" occurs in the morphospace and the region where "whorl overlap" is absent. What is "whorl overlap"? Consider the two hypothetical univalve shells given in figure 4.4. The two shells have identical values of the parameters *D* and *T* and differ only in values of *W.* The resultant shell forms are quite different, however. In the upper shell, the rate of whorl expansion is so low that each successive whorl overlaps the previous whorl; that is, the inner margin of the generating curve overlaps the outer margin produced one revolution earlier. In terms of the vector notation used earlier, whorl overlap occurs when the magnitude of $v_0$ at a given point is less than the magnitude of $v_I$ one revolution later around

FIGURE 4.4. Two computer-produced hypothetical planispiral shell forms from the *W–D* face of the theoretical morphospace given in figure 4.3. The successive whorls overlap in the upper shell form, whereas there is no whorl overlap in the lower shell form. Whorl overlap is common in organisms with univalved shell forms, whereas whorl overlap is absent in organisms with bivalved shell forms. *From Raup 1966. Used with permission of SEPM—Society for Sedimentary Geology (formerly Society of Economic Paleontology and Mineralogy).*

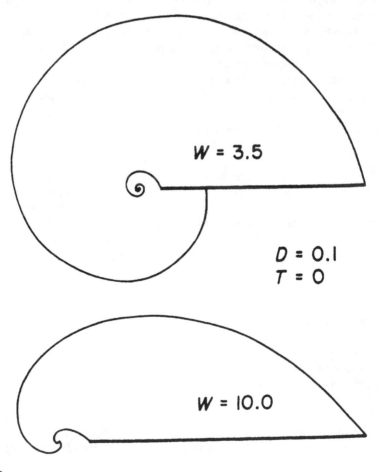

FIGURE 4.5. A diagrammatic representation of the theoretical morphospace given in figure 4.3, pointing out the surface boundary between the region of whorl overlap (upper part of the cube) and the region without whorl overlap (lower part of the cube) in the morphospace. *From Raup 1966. Used with permission of SEPM—Society for Sedimentary Geology (formerly Society of Economic Paleontology and Mineralogy).*

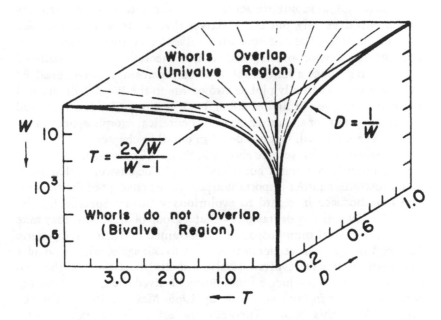

z-axis. In the lower shell in figure 4.4, the opposite is true, in that the rate of whorl expansion is so high that the inner margin of the generating curve never touches, much less overlaps, the outer margin produced one revolution earlier. That is, in any given revolution around the z-axis, the magnitude of $v_o$ is never less than the magnitude of $v_I$.

The two shells shown in figure 4.4 are planispiral; that is, they occur on the W–D face of the W–D–T theoretical morphospace, where T is equal to zero (figure 4.3). As one would expect, whorl overlap occurs in helicospiral shells as well, where T is not equal to zero. The boundary in the W–D–T theoretical morphospace between the whorl overlap region and the region of no whorl overlap is given in figure 4.5. The surface of this boundary is a function of the relative values of the three geometric parameters W, D, and T, as indicated in figure 4.5.

In terms of geometry, note that the whorl overlap boundary surface in the theoretical morphospace is similar to a shallow cone that has been quartered. This quarter-cone surface slopes downward from all

75

sides of the "Cube" toward the lower right front corner, that is, where the magnitudes of $D$ and $T$ are approaching zero and the magnitude of $W$ is increasing exponentially (figure 4.5). Above this surface, whorl overlap occurs. Below this surface, it does not. Raup (1966) designated the region above the surface as the "univalve region" of the theoretical morphospace and the region below the surface as the "bivalve region" (see figure 4.5). In geometric actuality, the theoretical morphospace is one of univalves only. But the univalves that occur below the surface may be considered as the separated valves of bivalved organisms in that they lack whorl overlap and so may theoretically be articulated together to produce a functional bivalved organism. A single bivalved organism would actually plot as two points in the $W$–$D$–$T$ theoretical morphospace given in figures 4.3 and 4.5, one point for one valve and a separate point for the other. Actual theoretical morphospaces of bivalved organisms will be considered in detail in chapter 5.

The whorl overlap surface shown in figure 4.5 was earlier described as a "geometric constraint boundary." Why? The answer to that question illustrates another important aspect of the concept of the theoretical morphospace in regard to evolutionary theory. Specifically, the evolutionary pathway or trajectory that a group of organisms may take in the theoretical morphospace is constrained by their phylogenetic history. I imagine the reader at this point to ask again, why? Consider organisms that have evolved bivalved shells. To articulate the two valves together to produce a functioning bivalved shell, *whorl overlap must be absent in both valves* (Raup 1966, McGhee 1980b; also see chapter 5 in this book). Theoretically, all evolutionary pathways through the morphospace below the whorl overlap surface in figure 4.5 are available for the separate valves of bivalved organisms, but no evolutionary pathway may ever rise above this surface (or even touch it; see McGhee 1980b). The whorl overlap surface in the morphospace thus acts as a barrier for the evolution of valve geometries in bivalved organisms.

Now consider organisms that have evolved univalved shells. Contrary to what we see for the evolution of bivalved valve geometries, the whorl overlap surface *does not act as a barrier for the evolution of univalved organisms*. Although Raup (1966) designates the volume above the whorl overlap surface as the "univalve region" (figure 4.5), the univalved organisms are in fact not confined to this region of the morphospace, although most univalves do indeed occur here. Univalve geometries that occur precisely on the whorl overlap surface have successive whorls that just touch one another and univalve geometries that occur under this surface have whorls that no longer touch. The geometric condition of having whorls that just touch one another is the common condition of advolute ectocochleate cephalopods and advolute gastropods. The more extreme condition in which the whorls have "uncoiled" and no longer touch one another is likewise still found in nature: the gyrocone, evolute, and heteromorph ammonoids immediately come to mind, as well as the evolute and vermetiform gastropods.

The geometric constraint boundary of the whorl overlap surface in figure 4.5 thus has the peculiarity of being an evolutionary barrier in one direction (bivalved valve geometries may never rise above, or touch, this surface), whereas it is only an evolutionary filter in the other direction (most univalve geometries lie above, but many univalve geometries touch, and some even fall below, this surface). The $W$–$D$–$T$ theoretical morphospace thus allows us to depict the effect of the evolutionary constraints of phylogenetic history and morphogenesis, two important concepts in the school of constructional morphology (see chapter 1).

I have attempted in this discussion to point out the many significant features of Raup's (1966) $W$–$D$–$T$ theoretical morphospace in regard to evolutionary theory. Now let us spend a little time exploring some of the problems of the $W$–$D$–$T$ morphospace. In the discussion of Raup's definition of the parameters at the beginning of this chapter, I mentioned a dimensional problem. To reiterate, the $D$ parameter is taken as the magnitude ratio of the two vectors $\mathbf{v}_I$ and $\mathbf{v}_O$; the $T$ parameter is scaled as the rate of magnitude change in $\mathbf{v}_T$ relative to the rate of magnitude change in $\mathbf{v}_O$; and the $W$ parameter is defined as the rate of magnitude change in $\mathbf{v}_O$ with time. Since both $D$ and $T$ are scaled with reference to $\mathbf{v}_O$, they both are therefore functions of $W$. The parameter $D$ is a direct function of $W$, and the parameter $T$ is an inverse function of $W$.

Why is this a problem? The problem is that $D$ and $T$ are not algebraically independent parameters of $W$. Variation in $W$ automatically produces variation in $D$ and $T$. Raup (1966) plotted the $W$–$D$–$T$ dimensions of the "Cube" at right angles to each other (see figure 4.3), which is indeed the desired orientation in euclidean space. In euclidean space, all dimensions are independent of one another; that is, variation along the $x$-axis produces no variation along the $z$-axis or $y$-axis. Since the condition of independent dimensionality is not met by Raup's parameters, they cannot properly be plotted as orthogonal to one another. Because variation in one parameter produces variation in another, the dimensional axes of the $W–D–T$ morphospace are noneuclidean and should be oriented at nonorthogonal angles to one another. The famous "Cube" is thus not a cube at all.

When we plot data points in a euclidean hyperspace, we do so with the knowledge that the dimensions of the hyperspace are independent, orthogonal, uncorrelated. If we then notice a correlation in the distribution of those points in the hyperspace, we can be certain that the correlation is due to the data themselves and not to the dimensionality of the hyperspace. Unfortunately, in the $W–D–T$ morphospace, we do not have this certainty. That is, if we notice a correlation between data points plotted in the $W–D–T$ morphospace, it may indeed be due to the data themselves, *but it may also be due to the fact that the dimensions of the morphospace are correlated with one another.* This point was forcefully pointed out in an insightful paper by Schindel (1990:273) who stated, concerning the empty areas in the $W–D–T$ morphospace (see figure 4.3): "I argue that the apparent proportion of unoccupied morphospace results more from Raup's algebra than from the biology of taxa with coiled shells."

The analysis of the distribution of actual measurement data in the $W–D–T$ morphospace must therefore be conducted cautiously. Particularly suspect would be the biological significance of any observed positive correlations between $D$ and $W$, and negative correlations between $T$ and $W$, as these correlations may be more "dimensional artifacts" than real.

Another criticism of the $W–D–T$ morphospace is that it contains only morphologies that are isometric. Gould (1991:421), for example, states: "Raup (1965, 1966, 1967, 1968) attracted much attention and inspired much fruitful work by defining a simple morphospace for coiled shells" but then adds the caveat "based admittedly on the unrealistic premises of isometric growth." Similar criticisms have been voiced by morphometricians (e.g., Goodfriend 1983) concerned with the accurate measurement of form. Raup was quite aware of this shortcoming in his model and explicitly included a computer simulation of a gastropod that exhibits ontogenetic changes in $T$ in Raup (1966, his Text-figure 5), and an analysis of actual anisometric growth in the ammonite *Paracravenoceras ozarkense* in the $W–D–T$ morphospace in Raup (1967, his Text-figure 11). But even these efforts have been criticized, for example, by Stone (1996b:907), who writes, concerning the gastropod simulation, that Raup's "parameters are constants and therefore have to be altered in an ad hoc manner throughout construction to produce anisometric form (Raup 1966)."

Much of the criticism leveled at Raup on the subject of anisometric growth I consider to be unfair and stems from the confusion between the two separate (but related) goals of theoretical morphology: the creation of theoretical *morphospaces* and the creation of theoretical models of *morphogenesis*. A theoretical model of morphogenesis should be as realistic as possible, and because the great majority of growth in nature is anisometric, Raup's (1966) model is therefore totally unrealistic as a model of morphogenesis. But a theoretical morphospace should be as morphologically complex as possible, *with the minimum number of dimensional parameters*. The surprising complexity of form seen in Raup's theoretical morphospace (figure 4.3) is created with just three parameters, and so as a morphospace, it is excellent. We shall return to the subject of anisometric growth in chapters 8 and 9, when we consider theoretical models of morphogenesis.

Returning to theoretical morphospaces, a very different type of theoretical model was used by Okamoto (1984, 1988a) and Ackerly (1989a) to create morphospaces of univalved shell form. Their respective models fall into the class of "generating-curve models with moving-reference frames" (see chapter 9). In Raup's model, changes that take place in the generating curve with time are always measured with reference to the $z$-axis, or coiling axis; hence this type of model is a "fixed-reference frame" model. The basic difference between a "fixed-reference frame" model and one with a "moving-reference frame" was perhaps best expressed by Okamoto (1988a:35) when he asked his readers to imagine

> a curved highway with a car moving at constant speed; this curve can be described using a fixed coordinate system, of course, but there is another method. The driver of the car must steer right or left to stay on the highway; if the steering operation is recorded exactly through time, the shape of the highway can be reconstructed by tracing the path of the car, without recourse to a fixed coordinate system.

Let us return to figure 4.1 at the beginning of the chapter. In addition to $v_I$, $v_O$, and $v_T$, this figure gives two vectors that have not yet been discussed. In the generating curve itself, we can specify a vector $v_R$ (for "radial vector") which extends from the center of the circular generating curve to its margin (see figure 4.1). Also extending from the center of the circular generating curve is the vector $v_D$ (for "directional vector") which points upward into the $y$-dimension, perpendicular to the page (i.e., perpendicular to the $x$-$z$ plane) in figure 4.1. We can use these two vectors to simulate the growth and expansion of the generating curve with time (using $v_R$), and the movement of the generating curve through space with time (using $v_D$), without any reference to a fixed coiling axis (the z-axis in figure 4.1).

Okamoto (1984, 1988a) used modified Frenet frames and differential geometry in developing his "growing-tube" model, which we will explore in chapter 9. But we can visualize what the model accomplishes by using the vectors $v_R$ and $v_D$. If the generating curve is to become larger with time, simulating the growth of a shell aperture, then the magnitude of $v_R$ must increase. If the generating curve is to move through space on a curved pathway, the direction of $v_D$ must change with time. And finally, $v_D$ may rotate about its own axis, causing the path of the generating curve through space to twist or revolve with time.

Okamoto (1988a) defined three parameters from his growing-tube model, which he then used to define the three dimensions of his theoretical morphospace:

1.  $E$ = the "radius enlarging ratio" of the growing tube. This parameter essentially measures the increase in the magnitude of $v_R$ with time.
2.  $C$ = the "standardized curvature" of the growing tube. This parameter essentially measures the change in direction of $v_D$ with time.
3.  $T$ = the "standardized torsion" of the growing tube. This parameter essentially measures the twist or revolution of $v_D$ with time.

The results of systematically varying the values of Okamoto's three parameters are illustrated in computer simulations of univalved shell morphology for various positions in the $E$–$C$–$T$ theoretical morphospace in figure 4.6. If all three parameters are zero, a simple cylindrical tube results (lower-facing corner of the morphospace in figure 4.6). That is, the generating curve moves through space with time but does not expand, curve, or twist. If $E$ and $C$ are zero but $T$ is not, a twisted cylindrical tube results (lower right corner of the morphospace). Note that the former straight line along the length of the simple cylindrical tube now traces a spiral around the tube (figure 4.6). If $E$ and $T$ are zero but $C$ is not, a curved cylindrical tube results. In the lower left corner of the $E$–$C$–$T$ theoretical morphospace (figure 4.6), Okamoto (1988a) gives a simulation for the condition in which the value of $C$ is 1.0, its maximum possible value. In that case, the cylindrical tube coils entirely back on itself, producing a torus. And finally, if $E$ is zero but $C$

**FIGURE 4.6.** Okamoto's *E–C–T* univalved theoretical morphospace (1988a). Computer-produced hypothetical shell forms illustrate the morphologic effects of varying the values of the geometric parameters *E*, the radius enlarging ratio (along the z-axis), *C*, the standardized curvature (along the y-axis), and *T*, the standardized torsion (along the x-axis). Compare the dimensional variations in form in this morphospace with that of Raup's "Cube" in figure 4.3. *Artwork courtesy of T. Okamoto. Reprinted with permission of the Palaeontological Association, London.*

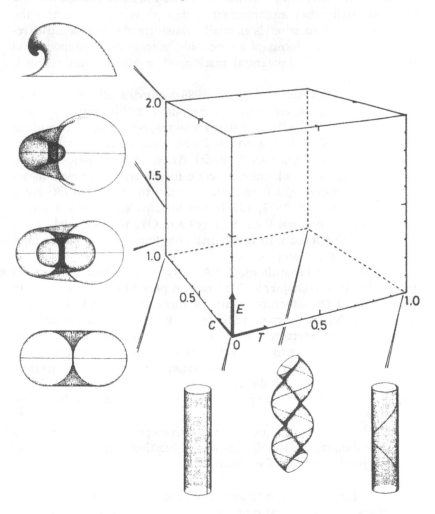

and $T$ are not, the cylindrical tube is twisted into a helix (lower rear corner of the morphospace, figure 4.6).

Further simulations are given along the left margin of the $E$–$C$–$T$ theoretical morphospace, where $T$ is zero but $E$ and $C$ are not (figure 4.6). All these simulations are planispiral and illustrate the morphologic effect of increasing the rate of radial expansion of the generating curve in the growing-tube model. The uppermost simulation is given in side view, rather than apertural view like the three simulations below it (actually, they are apertural *radiograph* views). Note how the torus (bottom simulation) is gradually transformed into geometries resembling the shell forms of ammonoids, planispiral gastropods, and the single valve of a potential brachiopod or bivalve mollusc shell (top simulation).

In contrast to Okamoto's (1984, 1988a) growing-tube model, Ackerly's (1989a) moving-reference frame model is a "kinematic" or "motion" model, analytically formalized with respect to the motion of the aperture of the shell during growth. Rather than the continuous differential equations of Okamoto's model, Ackerly's kinematic model is formulated in terms of finite, discrete increments of growth movement in the aperture, a formulation similar to the use of difference equations in Bayer's (1977, 1978b) and McGhee's (1978a) theoretical morphogenetic models. Both Ackerly's and Okamoto's models, however, make no reference to an external coordinate system in the morphogenetic simulation process.

A more detailed examination of Ackerly's (1989a) kinematic model will be deferred to chapter 9. Of interest at present is that each growth movement of the aperture occurs in three steps in the kinematic model, which here I generalize using the $v_R$ and $v_D$ notation outlined earlier. First, the aperture (i.e., the generating curve of other modelers) "translates" or moves forward a discrete distance in a direction specified by some angle relative to the direction of $v_D$. Second, the aperture "rotates" or changes the direction of $v_D$ by a discrete amount. Third, the aperture "dilates" or expands by a discrete increase in the magnitude of $v_R$.

Ackerly (1989a) uses these three growth steps to define two parameters that then define the dimensions of his theoretical morphospace of planispiral univalved shell form:

1. $\alpha$ = the "aperture rotation." The magnitude of the rotation is measured as the tangent angle to the spiral produced by the rotation in the simulation.

2. $\delta$ = the "aperture dilation." The magnitude of the dilation is measured as one-half the apical angle of the cone produced by the dilation in the simulation.

"Aperture translation" is not specified as a separate parameter; rather, the magnitude of the translation increment is used to scale the other two parameters. To simulate helicoconical geometries, two additional parameters are needed (see Ackerly 1989a):

3. $\gamma$ = the angle of the "downward" plunge of the aperture.
4. $\phi$ = the angle between the aperture and the rotation axis.

The theoretical morphospace of planispiral univalved shell form produced with the kinematic model is given in figure 4.7. The magnitudes of the two-dimensional parameters $\alpha$ and $\delta$ are measured in radians. The column of simulations at the far left shows the variety of conical forms produced by changes in the magnitude of dilation, from tall narrow cones (top left) to short broad cones (bottom left). Because the value of $\alpha$ is zero, there is no rotation, and straight cones are produced. The effect of changing magnitudes of rotation is illustrated by the remainder of the simulations, where the straight cones in the column of simulations on the far left of the morphospace are progressively more coiled with increasing magnitudes of $\alpha$, from left to right, in figure 4.7.

Okamoto's (1988a) and Ackerly's (1989a) theoretical morphospaces use parameters that make no reference to an external coordinate system. The simulated morphologies given in figures 4.6 and 4.7 appear to have coiling axes, but these axes are secondary by-products of the morphogenetic simulation process. This is not the case for the simulations in Raup's (1966) morphospace, where the coiling axis is specified at the beginning of the simulation and all subsequent growth in the generating curve is referenced in coordinates from this fixed axis. In addition, the models used by Okamoto and Ackerly to create their morphospaces were formally designed to allow for allometric growth, unlike the one used by Raup.

The dimensional parameters in Ackerly's (1989a) theoretical morphospace are also algebraically independent of one another. A specific goal of the kinematic model is the formal decoupling of translation, rotation, and dilation, unlike Raup's (1966) model, in which the parameter $W$ "reflects all three components of motion: translation, rotation, and dilation" (Ackerly 1989a:149).

FIGURE 4.7. Ackerly's α–δ planispiral univalved theoretical morphospace (1989a). Computer-produced hypothetical shell forms illustrate the morphologic effects of varying the values of the two geometric parameters α, aperture rotation (along the x-axis), and δ, aperture dilation (along the y-axis). Compare the dimensional variations in form in this morphospace with the *W–D* face of Raup's "Cube" in figure 4.3 (and the expanded view of that face given in figure 4.8). *Reprinted from* Paleobiology *and used with permission.*

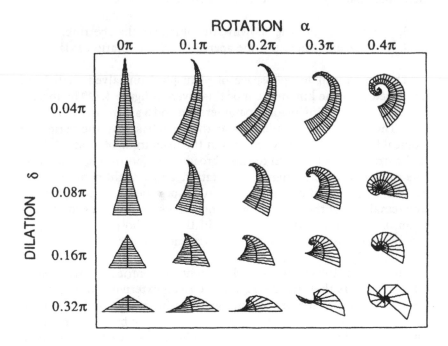

## THE EXPLORATION OF PLANISPIRAL
## UNIVALVED MORPHOSPACE

To date, no one has utilized the elegant theoretical morphospaces of either Okamoto (figure 4.6) or Ackerly (figure 4.7) to explore the existent distribution of form in nature. Graduate students looking for research topics take note! To begin exploring what morphology nature has actually produced in theoretical morphospace, we must return to Raup's (1966) *W–D–T* theoretical morphospace, given in figure 4.3. In figure 4.8 the *W–D* face of the *W–D–T* morphospace is given in more detail, showing the distribution of hypothetical planispiral shell forms that occur in this region of the morphospace. Note that all the shell forms in the upper left corner of the figure, above the line where *W* is equal to $1/D$, exhibit whorl overlap. Hypothetical shells that occur along the whorl overlap boundary line have whorls that touch but do not overlap. And shells that occur in the lower right corner of the figure have whorls do not touch at all, having open space between them.

All these planispiral shell forms are hypothetical and computer pro-

FIGURE 4.8. The computer-produced distribution of hypothetical plani-spiral shell forms from the $W$–$D$ face of the theoretical morphospace given in figure 4.3. Note the $W = 1/D$ curve delimiting the boundary between the region of whorl overlap (upper left of the figure) and the region without whorl overlap (lower right of the figure) in the morphospace. The hypothet-ical morphologies illustrated are commonly found among the coiled ceph-alopods. *From Raup 1967. Used with permission of SEPM—Society for Sedimentary Geology (formerly Society of Economic Paleontology and Mineralogy).*

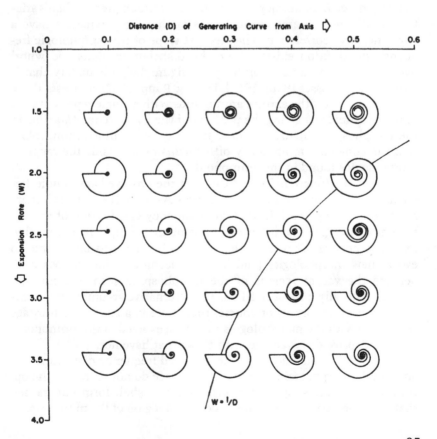

duced. Their resemblance to actual shells found in the ectocochleate cephalopods is obvious, however. In traditional cephalopod morphologic terminology, the hypothetical shell forms illustrated in figure 4.8 range, from the upper left to the lower right, from convolute to involute to advolute to evolute forms.

Where, then, do actual cephalopods occur in the theoretical morphospace? The frequency distribution of 405 ammonoid species, each representing a different genus, is given in figure 4.9. The peak of the distribution, showing the most frequently occurring morphology found in the ammonoids, is where the value of $D$ is around 3.5 and $W$ is around 2.1 (figure 4.9, cf. figure 4.8 for an illustration of shell forms that occur in this region of the morphospace). The frequency distribution is markedly asymmetrical, as can be seen in the pattern of the density contours. The distribution slopes sharply on either side of an elongate ridge that trends from left to right across the figure and that is oriented subparallel to the $D$-axis of the morphospace. Variation in the parameter $D$ in ammonoids is therefore much greater than variation in $W$, and the ammonoid values of the two parameters have a slight negative correlation. The steepest slope of the morphologic frequency distribution exists where the distribution abuts the whorl overlap boundary in the morphospace (figure 4.9), a boundary that it only slightly crosses (Raup 1967). Finally, Raup (1967) contrasts these measurements for the extinct ammonoids with a single representative of the nautiloids by giving the position of the living genus *Nautilus* in the morphospace (figure 4.9, cf. figure 4.8 for an illustration). Note that the convolute morphology of *Nautilus* falls outside the range of morphologies found in the ammonoids.

What is the functional significance of the observed ammonoid frequency distribution in the morphospace? As Raup (1967) noted, there might not be any at all, in that the frequency distribution of ammonoid morphologies that did evolve in time may simply represent phylogenetic chance. That is, if the ammonoids had had more time to evolve new morphologies (and had not become extinct 65 Ma ago), perhaps the vacant regions of the morphospace seen in figure 4.9 might eventually have been filled. The alternative argument from natural selection theory is, of course, that there is a adaptive advantage associated with the morphologies that have evolved in the ammonoids and an adaptive disadvantage with those that have not.

Raup (1967) concluded that the observed frequency distribution of ammonoid morphologies in the morphospace do not represent the optimization of any single functional aspect of shell form but, rather, that the geometries present in the occupied region of the morphospace

represent shell forms that minimize several different functional problems faced by the ammonoids. Ammonoids are generally confined to
the planispiral region, or *W–D* face, of the *W–D–T* theoretical morphospace, owing to streamlining and orientation stability constraints associated with their swimming mode of life. These swimming constraints also partially explain why the shells of most ammonoids have
whorls that contact and overlap one another, producing a solid disk
with no open spaces between the whorls. The location of the ammonoid frequency distribution to the left of the whorl overlap boundary

FIGURE 4.9. The frequency distribution of actual shell morphologies
found in 405 ammonoid genera, in the theoretical morphospace given in
figure 4.8. Contours measure the increase in density of genera per unit
area of the plot. The position of the living nautiloid genus *Nautilus* in the
morphospace is also indicated. *From Raup 1967. Used with permission of
SEPM—Society for Sedimentary Geology (formerly Society of Economic
Paleontology and Mineralogy).*

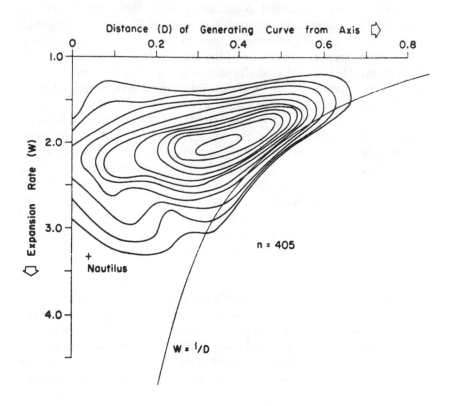

in figure 4.9 is also partially due to bioeconomical constraints in the efficient secretion of shell material by the animals (overlapped whorls use less shell material) and due to increased strength of the phragmocone for these types of shell (overlapped whorls are stronger).

The absence of ammonoids with $W$ values outside the range of 1.25 to 3.0 is argued to be largely due to the particular stability of shell geometries of these types (see figure 4.8 for an illustration of shell morphologies in these regions). Shell stability in floating and swimming is measured as the distance between the center of buoyancy and the center of mass for the shell divided by the diameter of the shell and is largely independent of $D$ but strongly a function of $W$ (Raup 1967). Shell geometries with $W$ less than 1.25 are very unstable, whereas those with $W$ greater than 3.0 are extremely stable. Raup concluded that ammonoids produced shells of an intermediate stability, balancing the need for attitude control against the need to change orientation in swimming, and so avoided the extrema of morphologies that are either too unstable or too stable.

Raup's (1967) theoretical conclusions were corroborated and expanded by Chamberlain's experimental work (1976, 1980). Using shell models with circular generating curves ($S$ equal to 1.0), Chamberlain (1976) experimentally determined the drag coefficients for a variety of shell forms. He found that shells with whorl overlap have much lower drag coefficients than those without. For shells with whorl overlap, he found two minima of drag coefficients: one at $W$ equal to 1.5 and $D$ equal to 0.1 and another at $W$ equal to 2.0 and $D$ equal to 0.43. Because hydrodynamic efficiency is the reciprocal of the drag coefficient, Chamberlain (1980) designated these two regions as "adaptive peaks" of high hydrodynamic efficiency in the $W-D$ theoretical morphospace (figure 4.10). In a subsample of actual ammonoids that have nearly circular whorl cross sections (to match the geometries of the experimental shells), he found that the frequency peak of actual ammonoid shells is centered about one of the two experimentally determined adaptive peaks (cf. figures 4.10 and 4.9), the one located at $W$ equal to 2.0 and $D$ equal to 0.43.

What about the other adaptive peak located at $W$ equal to 1.5 and $D$ equal to 0.1 (figure 4.10)? Chamberlain notes that this lower $W$ and $D$ region of the $W-D$ theoretical morphospace is one of lower static stability than the region with higher values of $W$ and $D$ (see the previous discussion of stability and Raup 1967). However, shell morphologies with very high stability ($W$ greater than 3.0) have very low hydrodynamic efficiency (high drag coefficients; see figure 4.10). Chamber-

**FIGURE 4.10.** The two X-marks in the *W–D* theoretical morphospace show the position of the two "adaptive peaks" of minimum drag coefficient (thus maximum hydrodynamic efficiency) determined by Chamberlain (1980). The numbered lines are contours of the drag coefficient (multiplied by 100), and the shaded areas show the distribution of ammonoids that have nearly circular whorl cross sections (generating curves) in the overall frequency distribution of ammonoid morphologies given in figure 4.9. The heavier shaded area shows the region where the greatest number of ammonoids occur (cf. figure 4.9). Note that this region is centered on one of the "adaptive peaks." *From Chamberlain 1980; copyright © Academic Press and used with permission.*

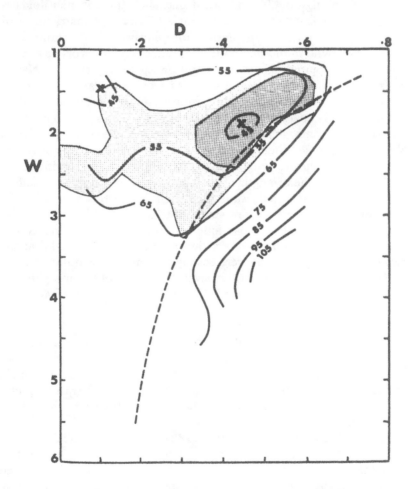

lain (1980) concluded that ammonoids produced shells whose geometry represents a trade-off between the antithetical needs of hydrodynamic efficiency versus stability.

Additional work with ectocochleate cephalopod morphologies with
noncircular whorl cross sections (Chamberlain and Weaver 1989) further confirmed the crucial role of stability for these jet-propulsion
swimmers.

Of all the ectocochleate cephalopods, Raup (1967), in his original
exploration of planispiral univalved morphospace, presented data for
only one nautiloid cephalopod, a species belonging to the modern genus *Nautilus* (figure 4.9), and noted that shell form in nautiloids is
quite different from that in ammonoids. The actual distribution of
nautiloid shell form in theoretical morphospace was examined by
Ward (1980), who found that nautiloids and ammonoids occupy virtually nonoverlapping regions in the morphospace (figure 4.11). Nautiloids do not show the range of diversity seen in the ammonoids and
are generally limited to strongly involute shell forms (cf. figures 4.11
and 4.8). Ward considered this limitation to be due in part to morphogenetic constraints. Nautiloids may be limited to simple globular to
slightly compressed shell shapes because of their simple nautiloid sep-
tum and septal suture, which provide little internal buttressing of the
shell, and thus the strength of the shell must be solely provided by
the curvature of the shell walls. The shell geometries present in the
nautiloids are generally more stable but have much lower hydrodynamic efficiencies than those found in ammonoids. The nautiloids
were therefore certainly slower swimmers, and Ward argued that the
two groups of cephalopods were ecologically quite dissimilar.

The possibility does exist, however, that some of the morphologic
dissimilarity between the nautiloids and ammonoids may have been
due to competitive displacement. Ward (1980) noted a very interesting
shift in morphologic frequencies in the nautiloids following the extinction of the ammonoids at the end of the Cretaceous, with postextinction nautiloids evolving new morphologies more similar to those
seen previously in the ammonoids. We will return to these animals
and examine this interesting evolutionary pattern further in chapter 7
when we consider the addition of the time dimension to theoretical
morphospaces.

As noted previously, Okamoto's (1988a) *E–C–T* theoretical morphospace and Ackerly's (1989a) α–δ theoretical morphospace have yet to
be utilized in exploring the theoretical range of planispiral univalved
shell form. The models behind those theoretical morphospaces have
been used to explore the bizarre morphology and extremely anisomet-

ric growth of heteromorph ammonites (Okamoto 1984; 1988a,b,c; 1993), and morphogenesis in the shells of monoplacophorans and nautiloids, and in the valves of brachiopods and bivalve molluscs, by Ackerly (1989a; 1990; 1992a). Thus measurements for the dimensional parameters of those two morphospaces have been obtained from actual organisms; they just have not yet been plotted in the morphospace formats of figures 4.6 and 4. 7. Why not? A comparative study of the

FIGURE 4.11. The frequency distribution of actual shell morphologies found in 72 Jurassic nautilids (solid contours) and 581 Jurassic ammonites (dashed contours), in the theoretical morphospace given in figure 4.9. Contours measure the increase in density of taxa per unit area of the plot. *From Ward 1980. Reprinted from* Paleobiology *and used with permission.*

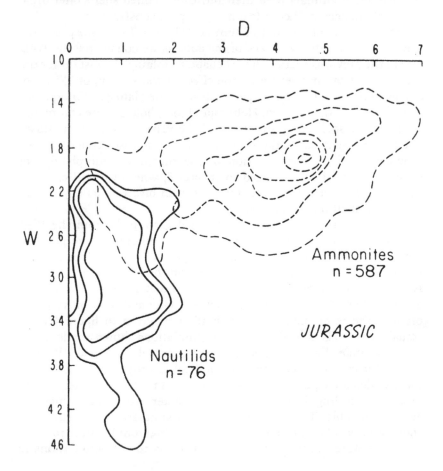

distribution of form in the $E$–$C$–$T$, $\alpha$–$\delta$, and $W$–$D$–$T$ theoretical morphospaces would make a very interesting research paper, given the different dimensionalities of the respective morphospaces. Such a study simply awaits an enterprising graduate student to pursue it.

## THE EXPLORATION OF HELICOSPIRAL
## UNIVALVED MORPHOSPACE

Snails are very interesting animals (they also are delicious with garlic butter!). We shall see in this section of the chapter that the geometry of their helicospiral shells is far from simple. Many other paleobiologists and biologists share my interest in snails, and so this section of the chapter is also a long one. I urge the reader to persevere, however, as these small animals with their intricately coiled shells offer illustrative challenges in theoretical morphospace construction.

Let us return once again to Raup's (1966) $W$–$D$–$T$ morphospace, this time to consider the $W$–$T$ face of the cube as we considered the $W$–$D$ face in the previous section of this chapter. Holding $D$ constant for the moment, we can now examine the effect of translation, or $T$, in producing helicospiral morphologies instead of the planispiral morphologies of the previous section. Helicospiral morphologies are commonly found in the shells of gastropods and the valves of bivalve molluscs (see figure 4.3). Because the shells of bivalve molluscs consist of two articulated valves, consideration of the theoretical morphology of these animals will be deferred to chapter 5, where we will consider the additional geometric constraints seen in bivalved (as opposed to univalved) shell morphologies.

Computer simulations of hypothetical gastropods from the $W$–$T$ face of the $W$–$D$–$T$ theoretical morphospace (figure 4.3) are given in figure 4.12. The horizontal axis in the figure is $T$, with values increasing from zero on the right to around 3.0 on the left. The vertical axis is $W$, with values increasing from around 1.5 at the top to around 10 at the bottom. Although produced by computer and thus entirely theoretical, the resemblance of the simulations given in figure 4.12 to actual gastropods is obvious. All the simulations shown are dextrally coiled, as indeed are the great majority of actual gastropods (Vermeij 1975). But computer-generated theoretical gastropods that are sinistrally coiled are equally easy to produce, and to the computer, the direction of coiling is just another parameter where all values are equally probable. The far right column of simulations in figure 4.12 shows a series of planispiral gastropods ranging from involute to advolute to evolute, from top to bottom. The top row of simulations in

FIGURE 4.12. The distribution of hypothetical helicospiral shell forms, produced by analog computer, from the *W–T* face of the theoretical morphospace given in figure 4.3. The horizontal axis in the figure is *T*, and the vertical axis is *W*. The hypothetical morphologies illustrated are commonly found among the gastropods. *Reprinted with permission from D. M. Raup and A. Michelson, "Theoretical morphology of the coiled shell," Science 147:1294–1295. Copyright © 1965, American Association for the Advancement of Science.*

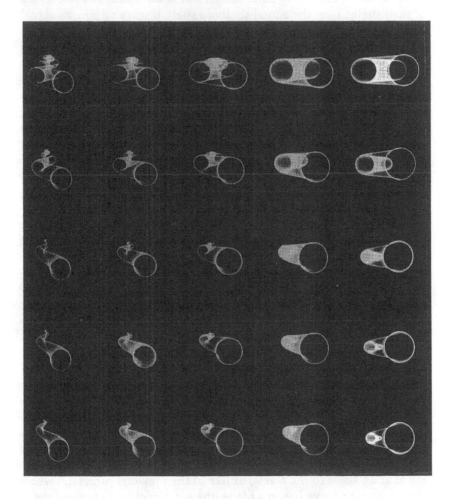

figure 4.12 shows a series of gastropods ranging from isostrophic (planispiral, with no spire) to orthostrophic (elevated spire). And in the lower left corner of the morphospace are illustrated hypothetical gastropods that have begun to "uncoil" in three dimensions, similar to vermetids.

There is, however, one major difference between the computer simulations given in figure 4.12 and actual gastropod shells. The simulations are more similar to axial sections of snail shells, or radiographs of snail shells, than the actual shell of a snail. If you examine an actual gastropod shell, you will see an enclosed helicocone with a single apertural opening, and the only way you can see the interior columnella and positions of the aperture in previous whorls is by sectioning the shell in half (or taking a radiograph of the shell). Thus, in all the simulations shown in figure 4.12, the aperture (the generating curve) lies in a plane that is radial to the coiling axis of the shell (as stated at the beginning of the chapter). Although some gastropods do indeed have shells of this morphology, most do not. In the majority of gastropods, the aperture lies in a plane that is oriented at some angle to the coiling axis.

Consider, as we already have done, a circular generating curve centered on the $x$-axis and displaced at some distance from the $z$-axis intercept (figure 4.1). The position of the inner margin of the generating curve is determined by $v_I$, and the outer margin by $v_O$. The generating curve thus initially lies in the $x$-$z$ plane, and in the computer simulations, this plane is rotated around the $z$-axis, the coiling axis, to produce the simulations given in figure 4.12, where in all cases, the generating curve lies in a plane radial to the $z$-axis. Gastropods that possess this type of shell were designated as having "radial apertures" by Linsley (1977).

Let us return to our circular generating curve, centered on the $x$-axis and displaced at some distance from the $z$-axis intercept, discussed in the preceding paragraph (figure 4.1). Now rotate the plane containing the generating curve about the $x$-axis itself. The margins of the generating curve lift out of the $x$-$z$ plane into the $y$-dimension— but not the innermost and outermost marginal points of the generating curve from the $z$-axis (i.e., $v_I$ and $v_O$ remain in the same position as before the rotation). The generating curve now lies in a plane oriented at an angle to the $z$-axis, unlike in the previous paragraph where the generating curve lies in a plane (the $x$-$z$ plane) that contains the $z$-axis. The magnitude of this angle is variable but has a theoretical upper limit of 90°. If the generating curve is rotated a full 90° about the $x$-axis, it will move from the $x$-$z$ plane into the $x$-$y$ plane, with the result that the aperture will be oriented perpendicular to the axis of coiling (the z-axis). Following the initial rotation of the generating curve about the x-axis, we can now rotate $v_I$ and $v_O$ around the z-axis to produce gastropods that have an aperture (generating curve) lying in a plane oriented at an angle to the coiling axis.

The angle between the plane containing the aperture and the coiling axis was called the "inclination angle" by Linsley (1977) and the "elevation angle" (of the coiling axis) by Vermeij (1971), and gastropods with this type of shell were designated as having "tangential apertures" by Linsley (1977). Raup's (1966) W–D–T morphospace does not contain this morphologic variable, and so to simulate gastropods with tangential apertures, an additional dimension would have to be added to Raup's initial four dimensions of W, D, T, and S. We shall return to this point later in this section when we consider actual utilization of the W–D–T morphospace by gastropods.

Theoretical shell forms with tangential apertures are produced in the E–C–T morphospace of Okamoto (1988a), and in the kinematic model of Ackerly (1989a). Computer simulations of hypothetical shell forms from the C–T face of the E–C–T morphospace are given in figure 4.13 for comparison with the W–T face of the W–D–T morphospace previously considered in figure 4.12. Okamoto arranged his morphospace dimensions so that the resulting simulated morphologies are similar to a mirror image of those produced in Raup's (1966) morphospace; that is, planispiral shell forms are located in the column of simulations on the left of figure 4.13, ranging from involute to evolute from top to bottom of the figure. Likewise, shell forms ranging from isostrophic to orthostrophic are arranged in the top row of figure 4.13 but are ordered from right to left.

Note now, however, the effect of Okamoto's parameter T, the "standardized torsion." All the simulations in the column on the far left of figure 4.13 have a T value of zero, and all have radial apertures. That is, an imaginary plane containing the generating curve would contain the apparent coiling axis of the theoretical shell form. The aperture is inclined and oriented at an angle to the apparent coiling axis in all of the other computer simulations, where T has a value greater than zero. Thus all the other theoretical shell forms have tangential apertures, as can be seen easily in the axial side view given for the simulation at C equal to 0.25 and T equal to 0.10, and also in the axial but more oblique view of the aperture given for the simulation at C equal to 0.75 and T equal to 0.20. (A brief aside: unfortunately, Linsley [1977] chose to illustrate the difference from his proposed "radial" and "tangential" apertures from an apical view and not an axial view. A precise

95

and unambiguous illustration of what constitutes an inclined ["tangential"] aperture may be found in Schindel [1990]. Viewed from an apical perspective [see Figure 1 in Linsley 1977], it is easy to confuse the geometric difference between a "radial" and "tangential" aperture with the difference between a "geometric generating curve" and a "biological generating curve" proposed by Raup [1966]. The latter difference results from a rotation quite different from the one needed to

**FIGURE 4.13.** The computer-produced distribution of hypothetical helicospiral shell forms from the *C–T* face of the theoretical morphospace given in figure 4.6. Compare with figure 4.12. *From Okamoto 1988a. Artwork courtesy of T. Okamoto. Reprinted with permission of the Palaeontological Association, London.*

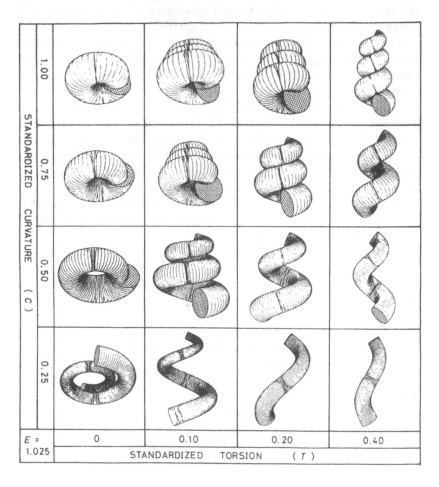

produce tangential apertures, a point that we will return to in chapter 5 when we consider "geometric" and "biological" generating curves in detail.)

Raup (1966) predicted, in the absence of any measurement data, that most gastropod shell forms would fall in the uppermost region of the $W$–$D$–$T$ morphospace, with values of $W$ ranging from 1.1 to 10, with values of $T$ ranging from zero to 4.0, and with values of $D$ ranging from zero to 0.5 (see figure 4.3). The functional significance of this entirely hypothetical (based on no measurement data) distribution of gastropod form was given a selectionist interpretation by Curry (1970) in his book *Animal Skeletons:*

> Snails, for instance, have only one valve and therefore if they had a large expansion rate the mouth of the shell would be very large in relation to the total body volume, leaving much of the body unprotected. (Limpets are an exception because they use the surface of the rock to which they cling as another valve). Snail shells often have a high translation rate because this is a convenient way of producing a globular shell, which is easier to carry around than a disc-shaped shell. (Curry 1970:14)

The first actual exploration of theoretical helicospiral univalved morphospace was conducted by Vermeij (1971), who studied the range in existent form found in 42 species of gastropods. He immediately encountered problems with Raup's morphospace. The parameter $T$, as defined by Raup (1966), could not be determined except by axial sectioning of the shell (thus destroying the shell). Vermeij (1971) used an alternative measure of $T$, one dating back to Thompson (1942): he measured $T$ as the cotangent of half the apical angle of the conical spire of the shell. This produces a static measure very similar to Raup's rate measure, in that the cotangent of half the apical angle is the ratio between the side of a right triangle adjacent to the apical angle ($v_T$) and the side opposite the apical angle ($v_0$). This measure of $T$ can easily be made with vernier calipers and does not destroy the shell. The parameter $S$ is of particular importance in gastropod morphology, as most snails have markedly noncircular apertures. Although this parameter is one of Raup's (1966) morphospace dimensions, he actually did not use it to produce the $W$–$D$–$T$ morphospace but most often used a simple circular generating curve ($S$ equal to 1.0) in theoretical simulations of gastropod shell form (as in figure 4.12). The parameter $D$ was of no use to Vermeij (1971), in that most of the snails he examined had $D$ values of zero. He thus proposed a new parameter, $E$, to replace

$D$, where $E$ is defined as "the smallest angle between the axis of coiling and the plane of the generating curve" (Vermeij 1971:16). This "elevation angle" of the coiling axis of the shell is the "inclination angle" of the aperture considered earlier. In short, all the snails under study had tangential apertures, and the angle of inclination of these tangential apertures was considered by Vermeij (1971) to be of major importance in the characterization of gastropod morphology. Finally, the parameter $W$ he chose to omit entirely, as being of less "interest in the properties and consequences of conispiral coiling" (Vermeij 1971:16), although he does use this parameter in later studies of gastropod morphology (Vermeij 1973a,b).

Of the four dimensions of the Raup (1966) morphospace, Vermeij (1971) used only two, $T$ and $S$, and had to design a new parameter $E$, which is absent in Raup's morphospace but is of great morphologic significance in the study of gastropod shell form. Vermeij (1971) made no attempt to plot his data in the $W$–$D$–$T$ morphospace, as in fact only one dimension, $T$, is present in the data. He does give all his measurements in tabular form, so one could theoretically design a three-dimensional $T$–$S$–$E$ space and plot the data in that alternative to Raup's $W$–$D$–$T$ morphospace. Graduate students note: construction of a $T$–$S$–$E$ theoretical morphospace, complete with simulated snail geometries, would make an interesting research paper. And empirical data already exist that could be plotted in the space.

Vermeij (1971) was able to group his data, based on their observed ranges of variation in the parameters $T$–$S$–$E$, into the Archaeogastropoda, "lower" Mesogastropoda, and "higher" Mesogastropoda plus Neogastropoda morphologic clusters. For example, he found that the archaeogastropods occur in morphospace regions of low values of $T$ but high values of $E$. The "higher" mesogastropods and neogastropods, in contrast, exhibit greater variability in $T$ but low values of $E$, and this same group exhibits the highest variability in $S$. Although not graphed explicitly, Vermeij did explore the distribution of gastropod form in theoretical morphospace. The next step in the procedure of theoretical morphology—that of the morphologic analysis of the significance of the distribution of real morphology in theoretical morphospace—he refused to take, stating only that "no attempt is made here to interpret the functional significance of variations in the several parameters" (Vermeij 1971:23)! This disclaimer is, in fact, not strictly true, as in the body of the text Vermeij makes several very interesting statements concerning the significance of the parameter $E$, such as "gastropods with low $E$ values have a far greater array of morphological possibilities open to them than do those with high $E$ values" and "lowering the axis of coiling has taken place many times in the subsequent evolution of the gastropods, and one may, in a strict geometrical sense, attribute much of the success of the group to this event" (Vermeij 1971:22). Yet it is true that his analysis of the distribution of existent gastropod morphology in T–S–E morphospace is very limited in comparison to the detailed analyses seen in the previous section of this chapter for the ammonoid and nautiloid molluscs.

In subsequent studies of gastropod morphology, Vermeij (1973a,b) uses Raup's parameter W, which he omitted in his 1971 study. In these two later papers, however, Vermeij (1973a,b) has in fact abandoned theoretical morphology in favor of morphometrics! He does continue to use the Raupian parameters W, T, and S, plus additional ones of his own, but he does so simply to quantify shell form in order to study the geographic and ecologic distribution of those quantifications. He plots, for example, the morphologic variables W and S versus the non-morphologic variable "species richness" (Vermeij 1973a:616–617), so it is clear that W and S are being used simply as morphometric variables. The use of Raup's theoretical morphology variables in morphometric studies—as opposed to theoretical morphologic studies—of gastropod morphology and ecology is likewise to be seen in the works of Davoli and Russo (1974), Graus (1974), Newkirk and Doyle (1975), Foote and Cowie (1988), and Johnston, Tabachnick, and Bookstein (1991). Graus (1974) explored latitudinal trends in gastropod morphology using a morphometric "form index" computed using Raup's W and T, and Davoli and Russo (1974) explored changes in W and T during both ontogeny and geologic time in five species of the gastropod Subula. Newkirk and Doyle (1975) used S, W, and T to quantify shell form in Littorina saxatilis in their genetic analysis of clinal variation in the morphology of that species; and Foote and Cowie (1988) used those same three parameters to examine the relationship between stabilizing selection and developmental constraint in shell form in the gastropod Theba pisana. Johnston, Tabachnick, and Bookstein (1991) used W to study ontogenetic changes in shell shape in Epitonium (Nitidiscala) tinctum, a gastropod that periodically develops prominent varices ideally suited to landmark-based morphometric analyses.

The next venture into the exploration of theoretical helicospiral univalved morphospace, following the initial study of Vermeij (1971), was taken by Kohn and Riggs (1975) in their study of seven species of the gastropod Conus. Like Vermeij, they immediately encountered problems with Raup's (1966) W–D–T morphospace. The parameter T, as formulated by Raup, could not be determined without taking axial sections of the shell, so Kohn and Riggs used the alternative measure of $T$ formulated by Vermeij. The parameter $D$ was generally equal to zero, so was of no use. The parameter $S$ Kohn and Riggs found to be highly variable and thus of more interest, so they substituted $S$ for $D$ to produce a $W$–$S$–$T$ morphospace instead of the usual $W$–$D$–$T$ morphospace. They then, however, added five parameters to those of $W$–$S$–$T$, parameters that they considered "necessitated by aspects of geometry specific to *Conus*" (Kohn and Riggs 1975:346), and these additional parameters were used in the morphometric section of the paper.

In terms of theoretical morphology, Kohn and Riggs (1975) found that species of the genus *Conus* exhibited only a small fraction of the range in $W$ predicted by Raup (1966) for the gastropods but at the same time exhibited values of $T$ exceeding the predicted range. Raup predicted that gastropods would occur in the $W$–$D$–$T$ morphospace in the region of $W$ ranging from 1.1 to 10, $T$ ranging from zero to 4.0, and $D$ ranging from zero to 0.5 (see figure 4.3). Kohn and Riggs report that *Conus* in fact occurs in the $W$ dimension from only 1.1 to 1.9 but in the $T$ dimension from 2.6 to 7.9. Thus the gastropod *Conus* occurs only in the very uppermost part of the predicted gastropod region in the $W$–$D$–$T$ morphospace (see figure 4.3), yet its distribution along the $T$ axis extends twice as far to the left as that shown in figure 4.3, where the $T$ axis is truncated at 4.0.

Because values of $D$ are zero in *Conus*, species of the genus do not extend into this third dimension of the $W$–$D$–$T$ morphospace (figure 4.3). In the $W$–$S$–$T$ morphospace, however, *Conus* ranges in the $S$ dimension from 5.6 to 19.0, the greatest range in magnitude variation seen for any of Raup's (1966) geometric parameters. Kohn and Riggs's (1975) research thus corroborates Vermeij's (1971) observation that the shape and inclination of the aperture are much more important geometric variables in gastropods than $W$ or $D$, which vary little. Likewise similar to Vermeij's study, they did not carry the theoretical morphologic analysis further by seeking to interpret the observed range of *Conus* in theoretical morphospace or to explore why *Conus* is confined to this region.

At about the same time as the Kohn and Riggs (1975) study, a foray into a quite different region of theoretical helicospiral univalved morphospace was being taken by Rex and Boss (1976), who chose to study gastropods that had become "open coiled." In gastropods of this type, the whorls are detached from one another, yet the shell still coils about an apparent coiling axis. This is in contrast to gastropods such as the vermetids that become uncoiled in ontogeny and thereafter exhibit irregular growth. In terms of the $W$–$D$–$T$ theoretical morphospace, open coiled gastropods would fall below the whorl overlap boundary in figure 4.5, as these morphologies have whorls that no longer overlap (or touch). Raup (1966) did predict that the gastropod morphologic range would extend below the boundary of whorl overlap in a very thin slice of the $W–D–T$ morphospace (see figure 4.3).

Rex and Boss (1976) found a range of variation in the values of the Raupian parameters $W$, $D$, and $T$ in the 15 species of recent open coiled gastropods under study, so they plotted the positions of these snails in the $W–D–T$ morphospace (figure 4.14). They were thus the first to be able to use directly Raup's (1966) unaltered $W–D–T$ morphospace in the actual exploration of the distribution of gastropod shell form. They did rotate the dimensions of the morphospace, however, and present a perspective entirely different from that given for the $W–D–T$ morphospace in figure 4. 3. In figure 4.3, the front upper right corner of the "Cube" is the position where the values of $W$, $D$, and $T$ all are zero, and the $T$-axis is the horizontal or $x$-dimension. In Rex and Boss's study, the $W–D–T$ (0–0–0) position is rotated to the rearmost lower corner, and the $T$-axis is rotated to form the vertical or $z$-dimension (figure 4.14). This rotation of perspective is necessary because their open coiled gastropods fall below the whorl overlap boundary surface (see figure 4.5). In the new rotation, the whorl overlap boundary surface is viewed from the underside (figure 4.14) rather than the topside (figure 4.5), as this is the region where the open coiled gastropod morphologies occur in the $W–D–T$ morphospace. Rex and Boss also plotted ecological, not just morphological, information in the theoretical morphospace (figure 4.14). They were primarily interested in seeing if the morphologies belonging to marine, fresh-water, and terrestrial snails would be distributed in different regions of the morphospace.

Similar to the Kohn and Riggs (1975) study of the single genus *Conus*, Rex and Boss (1976) found that the morphologic distribution of their open coiled species exceeded the $T$-dimensional limit of 4.0 predicted by Raup (1966) for the Gastropoda. In fact, more than half their sample (eight species) exceeded the $T$-value of 4.0, with two species ranging as high as 10.0 in the $T$-dimensional direction (figure 4.14). Despite this interesting result, Rex and Boss concluded that the morphologic distribution of these unusual open coiled gastropods did not "substantially increase" the amount of morphospace occupied by the Gastropoda, as Raup (1966) predicted. They noted that the condition of open coiling is extremely rare in gastropods and that the great majority of gastropods that do open coil also remain in close proximity to the whorl overlap boundary surface (figure 4.14).

Open coiled gastropod species are distributed all across the underside of the whorl overlap boundary surface, with three species oc-

**FIGURE 4.14.** The actual distribution of shell morphologies found in 15 species of open coiled gastropods, in the theoretical morphospace given in figure 4.3. Note that the perspective given in this figure represents a view from lower left back corner of the *W–D–T* cube given in figure 4.3 (and figure 4.5) and also where the cube has been rotated so that the *T*-axis is now vertical. The curved surface illustrated by gridlines in the figure is the underside of the boundary surface delimiting the region of whorl overlap given in figure 4.5. The numbers in squares, circles, and triangles represent gastropod species from fresh-water, marine, and terrestrial habitats, respectively. *From Rex and Boss 1976. Artwork courtesy of M. A. Rex.*

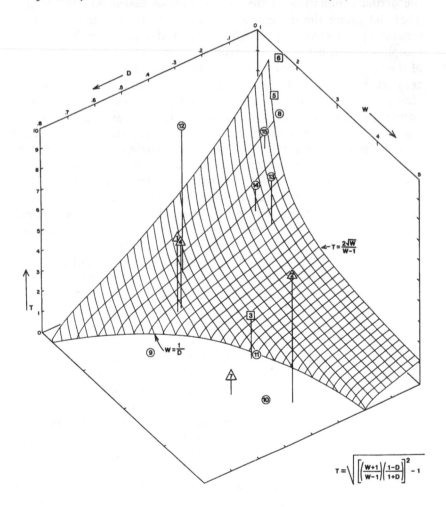

curring on the $W-T$ plane (where $D$ is zero) on one side of the boundary surface and three species on the $W-D$ plane (where $T$ is zero) on the other side of the boundary surface (figure 4.14). Thus there appears to be no single concentration of open coiled gastropod morphologies in the theoretical morphospace, a concentration that might suggest the presence of an "adaptive peak" of open coiled morphologies. There likewise appears to be no clustering of habitat, in that marine, freshwater, and terrestrial gastropods are scattered throughout the morphospace. Rex and Boss concluded that the condition of open coiling "approaches the limits of adaptation for most shelled gastropods" yet at the same time that "the selective significance of open coiling in any of the species treated here is very uncertain" (Rex and Boss 1976:294). But they did suggest that the condition of open coiling did not represent multiple convergences on a single morphology but that open coiling evolved independently in several different lineages to fill separate and dissimilar ecological niches.

Schindel (1990) was also able to use directly Raup's (1966) $W-D-T$ morphospace in his study of shell form in 53 modern species of gastropods, but at the same time he was highly critical of this theoretical morphospace and chose to design an alternative one of his own. Earlier in this chapter we noted a problem with the $W-D-T$ morphospace: the parameters of the morphospace are not algebraically independent of one another, and so the dimensions of the morphospace are not truly orthogonal. In his alternative theoretical morphospace, Schindel (1990) kept the parameter $W$ (he measured it differently, but it is essentially the same parameter as Raup's) but designed two new algebraically independent parameters to replace Raup's parameters $D$ and $T$. These new parameters Schindel designated as the "umbilical expansion rate," or $U$, and the "suture migration rate," or $M$.

The values of all the parameters in Schindel's (1990) study were obtained from the regression coefficients of measurements taken from gastropod shells plotted against whorl number, a modification of McGhee's mathematical technique (1980b:58-61, 76). Specifically, Schindel plotted the logarithms of measurements of "RP" ("radius from the coiling axis to the periphery of the shell," similar to $v_O$ in vector terms), "RU" ("umbilical radius," similar to $v_I$), and "SpHt" ("spire height," similar to $v_T$) against the whorl number for each gastropod shell. A polynomial function was then fit to these bivariate data plots by regression analysis (more discussion on why a polynomial function was chosen, rather than a linear function, is given in chapter 8 when we consider anisometric growth models). The first-order regression coefficients of the logarithms of RP, RU, and SpHt versus the whorl number yields the logarithms of the parameters $W$, $U$, and $M$, respectively.

In essence, Schindel's new parameters (1990) are simply measurement variants of the rate of magnitude change in the vectors $v_O$, $v_U$, and $v_T$ that we saw earlier (figure 4.1). If Raup (1966) had used these unscaled vector magnitudes to define his parameters $D$ and $T$, rather than scaling these parameters against the parameter $W$, the problem of parameter nonindependence would never have arisen.

Similar to the Rex and Boss (1976) study, Schindel (1990) chose to rotate the axes of the $W$–$D$–$T$ morphospace to a new orientation before plotting his gastropod data. Returning to figure 4.3, the front upper right corner of the "Cube" is the position where all the values of $W$, $D$, and $T$ are zero, and the $T$-axis is the horizontal or $x$-dimension. In figure 4.15A, Schindel (1990) rotated the axes of the morphospace twice: first, 90° about the $y$-dimension, so that the $W$–$D$–$T$ (0–0–0) position is now the front lower right corner rather than the front upper right corner, and, second, 45° about the $z$-dimension so that the $W$–$D$–$T$ (0–0–0) position is the right side lower corner rather than the front (in figure 4.14 this second rotation is 135° rather than 45°). As in the Rex and Boss study, the $T$-axis is now the vertical or $z$-dimension.

Schindel (1990) segregated his data into three "evolutionary grades" in order to plot them in the $W$–$D$–$T$ morphospace (figure 4.15A) and to contrast the distribution of those evolutionary grades in Raup's (1966) $W$–$D$–$T$ morphospace, versus their distribution in the alternative $W$–$U$–$M$ morphospace (figure 4.15B). The three evolutionary grades are archaeogastropods (species indicated by the numeral 1 in figure 4.15), nonsiphonate caenogastropods (the mesogastropods, indicated by the numeral 2 in figure 4.15), and siphonate caenogastropods (the neogastropods but represented in the study only by the cerithiids, indicated by the numeral 3 in figure 4.15). In figure 4.15A, a warp in vertical dimension is indicated in the plot of the morphospace position of the nonsiphonate caenogastropod *Truncatella valida*, as this species was

FIGURE 4.15. The actual distribution of shell morphologies found in 53 species of gastropods in Raup's $W$–$D$–$T$ theoretical morphospace (figure 4.15A) and the distribution of those same species in Schindel's $W$–$U$–$M$ theoretical morphospace (figure 4.15B). Note that the $W$–$D$–$T$ morphospace has been rotated so that the $W$–$D$–$T$ (0–0–0) position is now in the right-side lower corner, rather than the front upper right corner, as in figure 4.3. *From Schindel 1990. Copyright © 1990 by the University of Chicago and used with permission.*

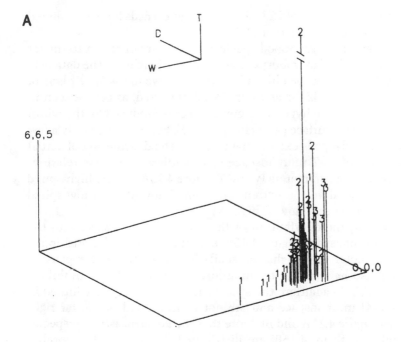

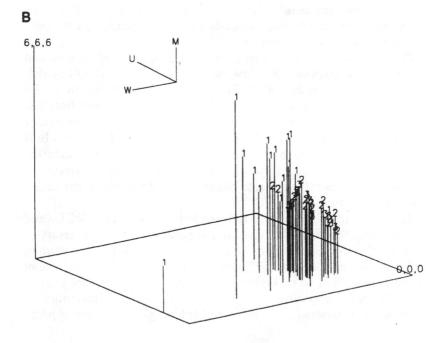

found to have a $T$ value of 22.9, a species that exceeds Raup's predicted maximum $T$ value by a factor of 5.7!

Note that all the gastropod species plot in a small region to the far right in the $W$–$D$–$T$ morphospace and that the heights of the data bars in figure 4.15A are shaped like a hyperbolic curve in the $W$–$T$ plane of the morphospace. This is as Raup (1966) predicted, as can be seen in figure 4.3, where the hyperbolic curve effect is produced by the whorl overlap boundary surface (see also figure 4.5), beyond which only a few open coiled gastropods extend (figure 4.14). The distribution of actual snails in the $W$–$D$–$T$ morphospace thus exhibits an inverse relationship between the parameters $W$ and $T$ (figure 4.15A), where high-spired shells are located at one extremum (high $T$, low $W$) and low-spired shells are at the other (low $T$, high $W$).

In contrast, the distribution of the same gastropod species in the $W$–$U$–$M$ morphospace (figure 4.15B) exhibits a direct relationship between parameter values; that is, snails having high whorl expansion rates ($W$) also have high rates of suture migration ($M$) and umbilical expansion ($U$). In addition, the species are more widely distributed in the $W$–$U$–$M$ morphospace and are not concentrated in the far right corner (cf. figure 4.15A and B). Note that the archaeogastropod species (numeral 1 in figure 4.15B) are distributed throughout the morphospace, having the full range of $W$, $U$, and $M$ rates. The nonsiphonate caenogastropods (numeral 2) have low to moderate rates in $W$, $U$, and $M$, and the siphonate caenogastropods (numeral 3) generally have only low rates of all three parameters.

The siphonate caenogastropods have high-spired shell forms and thus have high rates of $T$ but low rates of $W$ in Raup's (1966) model, as can be seen in figure 4.15A. These same gastropods occur in the low-rate region in Schindel's (1990) $W$–$U$–$M$ theoretical morphospace, because his parameter $M$, the suture migration rate and replacement for Raup's (1966) parameter $T$, is now independent of $W$, the whorl expansion rate. Thus high-spired shells now show their true architectural or morphogenetic nature: they have *slow rates in all aspects of expansion*, whether it in the rate of expansion of the whorl, the umbilicus, or the suture.

Even with the algebraically independent parameters of $W$, $U$, and $M$, Schindel (1990) found that additional parameters were necessary to examine the theoretical range of gastropod shell form. Like Vermeij (1971) and Kohn and Riggs (1975) before him, he found the aperture of the gastropod shell to be highly variable and hence of major significance. He thus added three new dimensions to his theoretical morphospace: the parameters apertural shape ($ApSh$), apertural angle ($ApAn$), and inclination angle (*InAn*). In fact, however, *ApSh* is the same as Raup's (1966) parameter *S* (shape of the generating curve), particularly because Raup left up the formal definition of this parameter to the individual researcher. The parameter *InAn* is the same as Vermeij's (1971) parameter *E* (elevation angle). The parameter *ApAn* is defined as "the angle between the axis of coiling and the long axis of the aperture as defined by the point of adherence and its opposite point, as marked in apertural view" ( Schindel 1990:295).

One of the main conclusions of Schindel's (1990) study is that high-spired gastropod shells are geometrically constrained to the low *W*, low *M*, and low *InAn* regions of the *W–U–M–ApSh–ApAn–InAn* morphospace, a conclusion that further corroborates Vermeij's (1971) and Linsley's (1977) thesis that these same shell forms have apertures that are more elongate and less inclined than in gastropods with low-spired shells. That is, Schindel (1990) found a positive correlation between the rates of *W* and *M* and of the angle *InAn* in gastropods, and he further suggested that high-spired shell geometries generally have a reduced range of morphogenetic variability available to them.

Other gastropod workers have abandoned the *W–D–T* theoretical morphospace entirely and seek neither to add new dimensional parameters to it nor to modify its existing ones. Cain (1977), for example, constructed a theoretical morphospace using the gastropod "spire index," which is defined as the height of the spire of the shell divided by the maximum diameter of the shell. Cain (1977:380) acknowledges that such a morphologic metric "is a crude measure of shape in gastropods (as compared with the various parameters defined by Raup in a series of papers, 1961–7)" but at the same time justifies its use because it is very simple to compute and the required measurements can be taken directly from photographs of shells in axial view. The latter is an important point, for much of the success of the theoretical morphologic analyses of ammonoids and nautiloids in planispiral univalved morphospace lies in the fact that the measurements required may be taken directly from photographs. I find it disturbing, however, that Cain felt compelled to apologize for the simplicity of the spire-index theoretical morphospace. This apology reflects the unfortunate misconception concerning theoretical morphology we examined earlier in chapter 2, namely, that all theoretical morphospaces must be the products of complex mathematics and sophisticated computer graphics. That is simply not true, and one of the goals of this book is to dispel that misconception once and for all.

Cain (1977) constructed a very simple two-dimensional morphospace using spire height and maximum diameter as parameters. In fact, his theoretical morphospace is very similar to the "theoretical morphospace of hypothetical triangular form" discussed in chapter 2 (see figure 2.4). Cain is, in essence, treating gastropod shells as if they were simple cones, with some snails having shells like tall narrow cones and others having shells like low broad cones. The morphospace of triangular form (figure 2.4) can easily be transformed into Cain's morphospace of conical form by replacing the parameter base width in figure 2.4 with the parameter maximum diameter, as the parameters apex height and spire height are geometrically the same.

Due to the ease of taking the necessary measurements from photographs, Cain (1977) was able to use as a database the major monographic literature devoted to gastropods. The study encompasses an enormous number of terrestrial species (the main thrust of the study), as well as the marine archaeogastropod, mesogastropod, and neogastropod groups, which have been the focus of the studies considered earlier in this section of the chapter. Cain also extended the study back in geologic time to consider extinct gastropod species. Results from that study will be discussed in chapter 7, where we shall look at the dimension of time from a theoretical morphologic perspective.

In the morphospace of conical form, Cain (1977) found major convergences on similar morphologies by widely different taxonomic groups. In addition, the same region in the morphospace is occupied by gastropod groups from different geographic areas around the world, whereas in the same geographic area the sympatric gastropod groups occupy different regions of the morphospace. Specifically, the fully terrestrial, free-crawling stylommatophoran snails in Europe, North America, Caribbean, and southwest Pacific islands all exhibit a bimodal density distribution in the morphospace, one mode with a spire index of about 3.0 (high-spired forms) and another with an index of about 0.5 (globular to trochoid forms). In every geographic area, there is a marked absence of snails with shells whose spire heights are the same as their maximum diameters. Only in the tropics, and in a few snail groups in the Philippines and New Guinea, were gastropods found with a unimodal distribution around a spire index of 1.0 and a density spread in the morphospace that bridges the bimodal gap between the high-spired and low-spired forms seen elsewhere (Cain 1978a,b).

Cain (1977) concluded that the multiple convergence of different taxonomic groups of snails on a limited (bimodal) region in the theoretical morphospace of conical form is due to the mechanical characteristics of the shell when considered as an object to be transported by the gastropod and that this in turn is related to the position in which the shell is carried by the crawling snail in its preferred habitat. In a mechanical analysis of shell stability, Cain further predicted that gastropods with high-spired shells should have a narrow foot, whereas the reverse should be true for gastropods with globular shells. The filling of the empty region in the morphospace by tropical rain forest faunas is attributed to the wider range of ecological niches available in such habitats, which would permit a greater range of shell geometries.

In a subsequent study, Cain (1980) added the parameter "whorl number" to his two previous parameters, spire height and maximum diameter. The conclusion of that later morphologic study is that in "very slender high-spired shells a high number of whorls is necessary for mechanical strength" and that a minimum limit exists for whorl cross section in even very small gastropod shells. The former conclusion was echoed a decade later by Schindel (1990) who, using a very different type of theoretical morphospace, concluded that the range of geometries available for high-spired gastropods was quite restricted.

Ekaratne and Crisp (1983) also chose to abandon Raup's (1966) parameters in a reformulation of Moseley's (1842) geometric model of gastropod form. They designed three alternative parameters and a "shell conversion factor" (which contains the three parameters) in their study of shell form in gastropods. The new parameters are $\lambda$, the ratio between diameters of successive whorls; $\beta$, the semiapical angle; and $\rho$, the ratio of aperture length and breadth in the apertural plane passing through the axis. Later in the paper, however, Ekaratne and Crisp (1983:793) acknowledged that $\lambda$ is equivalent to $W$, the cotangent of $\beta$ is equivalent to $T$, and $1/\rho$ is equivalent to $S$ in Raup's (1966) model.

Heath (1985) acknowledged Raup's model but also noted that not much attention had been paid to the extent to which the whorls in a gastropod shell may overlap. Heath then developed a morphospace in which the degree of whorl overlap may be examined. In the morphospace, an optimum degree of overlap was determined, one that minimizes the ratio of shell surface area to its internal volume. Surprisingly, in the subsequent exploration of this morphospace with measurements from actual gastropods, Heath found that all have degrees of overlap that are always greater than the optimum. The conclusion of the study is that some other factor than selection for economy is operating in determining shell form in the gastropods studied.

In summary of this section of the chapter, the exploration of helicospiral univalved morphospace has been much less comprehensive and straightforward than the exploration of planispiral univalved morphospace, even though more researchers have been involved. The use

**109**

of Raup's $W$–$D$–$T$ theoretical morphospace for planispiral organisms was facilitated by the fact that in most cases, the measurements needed from planispiral organisms could be obtained directly from photographs. Hence the success and speed in which actual shell form in ammonoid and nautiloid cephalopods was analyzed in a theoretical morphologic context, as discussed earlier in this chapter.

Unfortunately, the parameters of the $W$–$D$–$T$ morphospace are not so easily obtained for helicospiral organisms. For gastropods in particular, determining these parameters usually involves axial sectioning of the shell, as was done in the studies by Davoli and Russo (1974) and Graus (1974). Alternatively, Raup's (1966) parameters can be modified or redefined to facilitate their measurement. Vermeij (1971; 1973a,b) redefined both $T$ and $W$ to allow for the measurement of these parameters from actual specimens using vernier calipers, making unnecessary the destruction of the shell through axial sectioning. These modified parameters were also used by Kohn and Riggs (1975) in measurements of apertural view photographs in their study of shell form in *Conus*.

Schindel (1990:275) also makes the point that the parameters of the $W$–$D$–$T$ morphospace often yield a "lack of satisfying results" in the analysis of gastropod shell form. That is, its parameters may show little or no variability in actual gastropod populations. As discussed earlier in this section of the chapter, many gastropod morphologists soon found the parameter $D$ to be of little use because it is most often equal to zero in gastropod shells. Most found the different angular orientations of the aperture relative to the rest of the shell to be of major significance (Vermeij 1971, Kohn and Riggs 1975, Schindel 1990; see also McNair et al. 1981), necessitating the creation of new parameters to model these attributes of gastropod form. Others attempted to make the parameter $S$ more biologically realistic, in that it is generally modeled as a circle or an ellipse, both of which are rarely found in actual gastropod apertures. Harasewych (1982) designed a much more complex model for $S$, one in which a siphonal canal is also included. Hallers-Tjabbes (1979) considered even more dimensional complexity in modeling the actual aperture of *Buccinum undatum* but finally abandoned theoretical morphology in favor of morphometrics.

Difficulty in measurement of the model parameters is only partially a function of the model used; it is also a function of the geometry of the helicospiral shell itself (Verduin 1982). The model used by Raup (1966) falls in the general class of "generating-curve models with fixed-reference frames" (see chapter 9). Using the very different "moving-reference frame" class of model, Ackerly (1989a:158) still notes for helicospirals that "three-dimensional analyses require more geometric formality than their two-dimensional counterparts" and that "the key element of the kinematic analysis is defining criteria for fixing a coordinate system to the aperture," which is often not easy. The difficulty in measuring the parameters of the moving-reference frame model in helicospirals led him to design a new technique for analyzing the gastropod shell form—stereographic projection (Ackerly 1989b)—a technique that he describes in passing as "less rigorous, but still very informative" (Ackerly 1989a:159). Obtaining measurements for the algebraically independent parameters in Schindel's (1990) $W$–$U$–$M$ morphospace still requires axial sectioning of actual gastropod shells. In addition, in his study he outlines a meticulous and labor-intensive procedure, one requiring seven separate steps and taking a full half-page of text to describe, which must be repeated for each snail shell in order to obtain the necessary measurements (Schindel 1990: 282–283)! In fact, the extensive theoretical morphologic analysis of gastropod form conducted by Cain (1977; 1978a,b) was made possible by abandoning these more realistic morphospaces of helicospiral geometry. Cain chose instead to use a much less realistic morphospace, one in which gastropods are treated as if they are simple cones but also one in which the necessary measurements for the parameters could be taken from photographs.

Can a theoretical morphospace of helicospiral form be created that is both morphologically realistic and mensuratively simplistic? That would be an intellectually daunting challenge indeed, especially considering the numerous efforts seen in the previous studies of the geometry of helicospiral form that were discussed in this lengthy section of the chapter! Yet the challenge still exists and awaits someone to accept it.

# V

# Spirals and Shells II: Theoretical Morphospaces of Bivalved Accretionary Growth Systems

In this chapter we shall examine the various theoretical morphospaces that have been designed for bivalved organisms using accretionary growth systems, in contrast to chapter 4, in which we examined theoretical morphospaces for univalved organisms. Bivalved organisms must maintain two separate valves in particular orientations to each other in daily life and must continue to maintain or modify those orientations in time as the organism grows. The evolutionary constraints imposed on valve morphology in bivalved shells are not easily discerned in the univalved theoretical morphospaces discussed in chapter 4, where in every case the valves of an individual bivalved shell must be plotted as two separate points in the morphospace. In a bivalved theoretical morphospace, the morphology of each bivalved individual plots as the single unique point that this individual represents. In addition, the geometric characteristics of a bivalved morphospace are a bit more complex those of a univalved morphospace, as we shall see.

## THEORETICAL MORPHOSPACES
## OF BIVALVED SHELL FORM

The first computer-generated theoretical morphospace for bivalved organisms was created by McGhee (1978b, 1980b), using the geometric parameters $W$ and $D$ of Raup (1966) that we examined in chapter 4. The morphospace was specifically designed for organisms with planispiral shells; thus Raup's parameter $T$ could be omitted. The dimensions of the morphospace are determined by the geometric parameters of the two valves of the shell, however, and not just by the parameters in isolation. That is, the spatial dimensions of the morphospace are determined by placing the geometric parameters with reference to the first valve orthogonal to the geometric parameters with reference to the second valve. Because there are two valves in a bivalved shell, and two parameters $W$ and $D$, the resultant morphospace has four dimensions.

To simplify matters for visual presentation, a series of two-dimensional slices were taken through the four-dimensional theoretical morphospace and examined sequentially in order to explore the pattern of morphologic variation in the total four-dimensional hyperspace. One such two-dimensional slice is given in figure 5.1. Because the morphospace was created to analyze the evolution of shell form in the Brachiopoda, one valve is designated as "dorsal" and the other as "ventral," which is the anatomical condition for this group of animals. One could just as easily designate one valve as "right" (or dextral) and the other as "left" (or sinistral) if one is examining planispiral bivalve molluscs, and so on, depending on the organisms under study. As the value of $W$ or $D$ in one valve of a brachiopod shell is algebraically independent of the value of those same parameters in the other valve, the dimensions of the morphospace given in figure 5.1 are orthogonal and uncorrelated.

Most planispiral bivalve molluscs are equivalved; that is, the left and right valves of the shell are mirror images of each other (thus the values of $W$ and $D$ in the left and right valve are the same). The brachiopods, however, are inequivalved. The geometry of the dorsal and ventral valves of a brachiopod shell are not mirror images of each other, and the values of $W$ and $D$ in each valve are different. In figure 5.1, the two-dimensional slice was taken through that region of the four-dimensional morphospace where the value of $D$ on the dorsal-$D$ axis is 0.01 and the value of $D$ on the ventral-$D$ axis is 0. 1. Because the parameter values on those two dimensional axes are held constant, the remaining two dimensions of dorsal-$W$ and ventral-$W$ may be displayed in a simple two-dimensional diagram of hypothetical brachiopod shell form (figure 5.1). If one takes different values of $D$ along the dorsal-$D$ and ventral-$D$ axes of the four-dimensional morphospace, the resultant morphologies in the two-dimensional dorsal-$W$ and ventral-$W$ slice will be different from those shown in figure 5.1 (though actually not by much, unless very large values of $D$ are chosen). The values

113

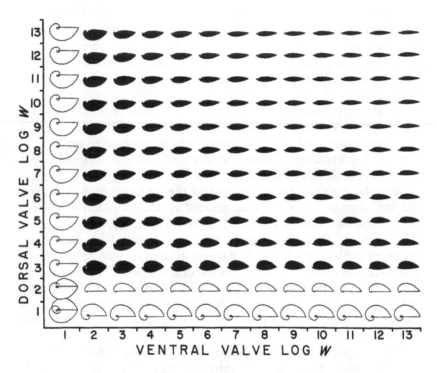

**FIGURE 5.1.** The dorsal-*W*–ventral-*W* slice through McGhee's four-dimensional bivalved theoretical morphospace (1980b), showing the potential spectrum of computer-produced hypothetical biconvex brachiopod shell morphologies. Axes measure differential valve convexity in terms of the geometric parameter *W*, the whorl expansion rate. Unshaded hypothetical dorsal and ventral valves shown immediately adjacent to the axes illustrate the morphospace region of "whorl overlap" in these valves, the region of the morphospace unavailable for bivalved shells (see the text for discussion). *From McGhee 1980b.*

of $D$ chosen to produce figure 5.1 were empirically determined, in that the majority of brachiopod shells have dorsal- and ventral-$D$ values, respectively, of 0.01 and 0.1 (McGhee 1980b).

Note that in figure 5.1 a series of hypothetical dorsal and ventral valves are shown as single valves and are unshaded, immediately adjacent to the morphospace axes. These valves have "whorl overlap," a condition defined in chapter 4 (see figure 4.4). Whorl overlap must be absent in both valves to articulate the valves together to form a single functioning bivalved shell (Raup 1966, McGhee 1980b). The region of whorl overlap shown in figure 5.1 is a function of the values of $D$ chosen on the other two dimensional axes of the morphospace. In the total four-dimensional theoretical morphospace, the region of whorl overlap expands away from the dorsal-$W$–ventral-$W$ axes as smaller values of $D$ are encountered on the dorsal-$D$–ventral-$D$ axes and contracts toward the dorsal-$W$–ventral-$W$ axes with larger values of $D$.

The unshaded single valves shown in figure 5.1 cannot be articulated together to form a functioning bivalved shell. Now examine the first shaded bivalved shell, with both a dorsal and a ventral valve, in the lower left corner of figure 5.1, at the position where the dorsal valve has a $W$ value of $10^3$ and the ventral valve has a $W$ value of $10^2$. These represent the minimum values of dorsal and ventral $W$ attainable in this region of the theoretical morphospace if one is to produce an articulated bivalved shell. These minima, and the whorl overlap region in general, constitute a major "geometric constraint boundary surface" in the four-dimensional bivalved morphospace, just like that in the three-dimensional $W$–$D$–$T$ univalved morphospace discussed in chapter 4 (see figure 4.5). The dorsal-$W$–ventral-$W$ morphospace slice given in figure 5.1 has proved to be very powerful in interpreting the evolution of shell morphology in the Brachiopoda, as will be explained later in this chapter when we consider actual utilizations of the bivalved theoretical morphospace.

There is yet another morphologic variable important to animals with shells consisting of two separate valves. That morphologic variable was outlined in a general way by Raup (1966), who discussed the possible condition of an "angle between the geometric and biological generating curves of a valve," as seen in figure 5.2. The variable was formalized by McGhee (1980b), who designated it as the parameter $B$. What is it, and what are "geometric" and "biological" generating curves? In all the computer simulations given in figure 4.8 for univalved shells and figure 5.1 for bivalved shells, the position of the generating curve is determined by the two vectors $v_I$ and $v_O$ in the $x$-$y$ plane, as set out in chapter 4 and illustrated in figure 4.1. In these

115

simulations, the two vectors always point in the same direction at any given point in time and determine what Raup (1966) elaborated as the "geometric generating curve," which until now we have been simply designating as the "generating curve." The geometric generating curve, as can be seen in figure 5.2, always lies in a plane radial to the coiling axis (the z-axis). We can observe, however, that many—particularly bivalved—animals possess valves in which the aperture or commissure of the valve does not lie in a plane radial to the coiling axis, and this nonradial orientation was termed the "biological generating curve" by Raup (1966). In vector terms, the vectors $v_i$ and $v_o$ do not point in the same direction at some given point in time for a biological generating curve, but rather, there is an angle between them.

An alternative way to visualize Raup's (1966) "angle between the geometric and biological generating curves of a valve" or McGhee's (1980b) $B$ is to think in terms of rotations. Let us return to figure 4.1 and consider a circular generating curve lying in the x-z plane, centered on the x-axis, and displaced at some distance from the z-axis intercept. If the initial geometric generating curve, lying in the x-z

FIGURE 5.2. Illustration of the difference between the "geometric generating curve" (which always lies in a plane radial to the coiling axis) and the "biological generating curve" (which may not). The "angle between geometric and biological generating curves" was designated the parameter $B$ by McGhee (1980b). *From Raup 1966. Used with permission of SEPM— Society for Sedimentary Geology (formerly Society of Economic Paleontology and Mineralogy).*

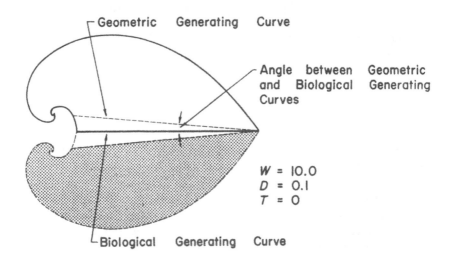

plane, is rotated about an axis located parallel to the $z$-axis but passing through the center of the generating curve, then the margins of the generating curve will move out of the $x$-$z$ plane into the $y$-dimension. The new rotated orientation produces a biological generating curve in that the inner and outer margins of the generating curve (i.e., $v_I$ and $v_o$) no longer lie in a single plane radial to the $z$-axis (the $x$-$z$ plane, in case of the initial generating curve).

What does this mean? Examine the lower valve given for the hypothetical bivalved animal in figure 5.2. The shaded portion of the valve shows the position of the geometric generating curve (also shown by the dashed line in the upper valve), which is a radius drawn from the coiling axis. However, the two valves actually join at a margin, the biological generating curve, which is at an angle to this radius. At the margin where the two valves actually join the inner vector $v_I$ is in advance of the outer vector $v_o$; that is, the two vectors no longer point in the same direction at the same point in time. The angle between the termini of $v_I$ and $v_o$ was designated $B$ by McGhee (1980b). If there is no angle between these vectors, then $B$ is equal to zero, and the geometric and biological generating curves are one and the same (as is the condition in the computer simulations given in figures 4.3 and 5.1). For a single valve, the values of $B$ are defined as positive when $v_I$ is in advance of $v_o$ at some point in time and negative when $v_o$ is in advance of $v_I$ (see also the discussion of $B$ sign determination in McGhee 1980b).

The theoretical spectrum of all possible $B$ angle combinations between two valves articulated into a single shell are given in a two-dimensional representation in figure 5.3. The distribution of values of the parameter $B$ in the natural world has not been examined as fully as that of the parameters $W$ and $D$, and so the parameter is similar to Raup's parameter $S$ (shape of the generating curve) for which the same is true. Raup's (1966) theoretical morphospace for univalved organisms actually was designed with four dimensions: $W$, $D$, $T$, and $S$, although most people are familiar with only the three dimensions of the famous "Cube" (figure 4.3). Likewise, McGhee's (1980b) theoretical bivalved morphospace was designed with six dimensions: $W$, $D$, and $B$ for the two valves in a bivalved shell, although most are familiar with only the dorsal-$W$–ventral-$W$ morphospace of figure 5. 1. The extra dimensions of dorsal $B$ and ventral $B$ do reveal interesting patterns in brachiopod evolution, however, and these will be discussed in the morphospace exploration section of this chapter.

There are some major problems or limitations with McGhee's (1980b) bivalved theoretical morphospace, however. The morphospace

**117**

was designed for (1) planispiral shells only and (2) biconvex shells only. Because the morphospace was created to analyze shell form in the Brachiopoda, the first limitation was not a problem, although it will be when analyzing shell form in the bivalve molluscs or ostracodes, and so on, as shell form in these organisms is not planispiral. This problem was tackled by Savazzi (1987), who produced a theoretical morphospace for helicospiral bivalved organisms, which we shall consider later in this section of the chapter.

The second limitation is a major one even for the Brachiopoda, however, and thus the early theoretical morphologic studies of shell form in these animals was confined to the biconvex forms (McGhee 1978b; 1979a,b; 1980b,c). The geometric problem of nonbiconvex shell form was examined in a series of models (McGhee 1980a, 1991, 1999d) that produced a new bivalved theoretical morphospace including the total spectrum of differential valve convexities, which we shall discuss next.

In early modeling, it became clear that to examine the theoretical morphology of nonbiconvex shell form, Raup's (1966) differential con-

FIGURE 5.3. The nine-cell spectrum of all possible permutations of sign in the parameter $B$ for hypothetical bivalved organisms. *From McGhee 1980b.*

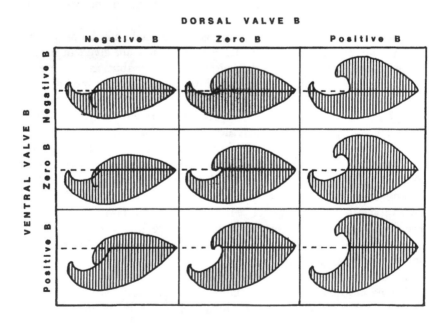

vexity parameter—the whorl expansion rate, or $W$—must be abandoned. Remember in chapter 4 we noted in general that the parameter $W$ was a measure of the rate of magnitude change in $v_0$ relative to its change in direction with time. At this point, I must deviate slightly from my stated goal at the beginning of chapter 4 to discuss models in general terms only and to defer analysis of their mathematics to chapters 8 and 9. Only a slight deviation, I promise, so the reader may cease groaning. Let us decompose the vector $v_0$ into its two components: its magnitude and its direction. The magnitude of the vector is the radial distance, which we shall designate as $r$, from the coiling axis to the outer margin of the spiral produced by the generating curve moving through space (illustrated in figure 8.1 if a graphic visualization is needed). The direction of the vector is measured by the angle of rotation, designated as $\theta$, of the generating curve around the coiling axis.

Initially, Raup (1966) defined $W$ as the ratio of radii $r$ after one full revolution about the coiling axis; that is, he specified the change in the coiling angle $\theta$ to be 360° (or $2\pi$ radians). Animals with shells having very low spiral curvatures, such as brachiopod valves, possess spirals that never coil through $2\pi$ revolutions, and to measure these whorl expansion rates, a more general definition of $W$ must be derived (McGhee 1980b):

$$W = (r_{\Delta\theta} / r_0)^{2\pi/\Delta\theta} \tag{5.1}$$

where again $r_0$ and $r_{\Delta\theta}$ are measured radii from the coiling axis to the outer margin of the spiral, and $\Delta\theta$ is the angle between the two measured radii being compared.

The spirals present in valves having flatter and flatter convexities have large radial expansions that occur with ever decreasing values of $\Delta\theta$, such that some very flat brachiopod valves have $W$ magnitudes of $10^{11}$ to $10^{12}$ (see figure 5.1). As $\Delta\theta$ approaches zero, the magnitude of $W$ approaches infinity; thus the parameter cannot be used to describe incremental growth that takes place entirely in a single plane (i.e., when there is no $\Delta\theta$). Yet such planar growth forms are common in the essentially flat, "lidlike" valves of some nonbiconvex brachiopods.

To overcome this difficulty while still maintaining the spiral geometric methodology, the parameter $\alpha$, the tangent angle of the spiral, may be used instead of $W$ as a measure of spiral curvature or valve convexity (again, see figure 8.1 if a graphic visualization is needed). Now let us consider the two possible end-member geometric extremes: "growth" that occurs as a circle in a plane and "growth" that occurs as a line in a plane. In the case of the perfect circle extreme, the tangent angle to the curvature of the "spiral" would be 90° (and $W$ is equal to 1.0, as the radii after one full revolution about the coiling axis are exactly the same length). In the case of the linear extreme, the tangent angle to the "curvature" of the "spiral" would be 0° (and $W$ would be undefined, as $\Delta\theta$ is equal to 0). The relationship between $\alpha$ and $W$ for normal spiral geometries between the two geometric extrema just discussed is given in the upper right quadrant of figure 5.4.

Last, to fully characterize nonbiconvex brachiopod shell form, we must consider the morphologic condition of concavity. Many nonbiconvex brachiopods have valves that are neither convex nor planar in form but that are actually concave. To address this condition, we can define convex valves as having positive values of $\alpha$ (and $W$) and concave valves as having negative values of $\alpha$ (and $W$). The relationship between $\alpha$ and $W$ for concave valves is given in the lower left quadrant of figure 5.4.

A new theoretical morphospace of planispiral bivalved shell form may now be created, one that includes shells composed of valves that may be convex, planar, or concave (figure 5.5). All possible permutations of these differential valve morphologies, in combination to produce a hypothetical bivalved shell, are represented in figure 5.5. Note that fully half of the possible valve combinations, those in the upper right triangle, are geometrically impossible shell forms. These permutations represent potential shell forms, for example, in which both the dorsal and ventral valves are concave—depicting, in essence, brachiopods that have been turned "inside out." Only those dorsal and ventral valve combinations that fall into the shaded region (lower left triangle in the figure) depict biologically feasible brachiopods. Even here, brachiopods cannot occupy the full spectrum of the lower left triangle, as whorl overlap must be absent in both the dorsal and ventral valve in order to articulate the two valves together to form a functioning shell (as pointed out earlier).

The distribution of distinctive brachiopod shell forms, using conventional brachiopod morphologic terminology, is given in the potential brachiopod morphospace in figure 5.6. Our previously considered biconvex brachiopod shell forms occur in the lower left corner of the geometric continuum. However, the new continuum now extends upward, along the $y$-axis of the morphospace, to include shell forms with truly planar dorsal valves (plano-convex shells) and farther upward to include those with concave dorsal valves (concavo-convex shells). Likewise, the new continuum extends along the $x$-axis of the theoretical morphospace to include shell forms with truly planar ventral

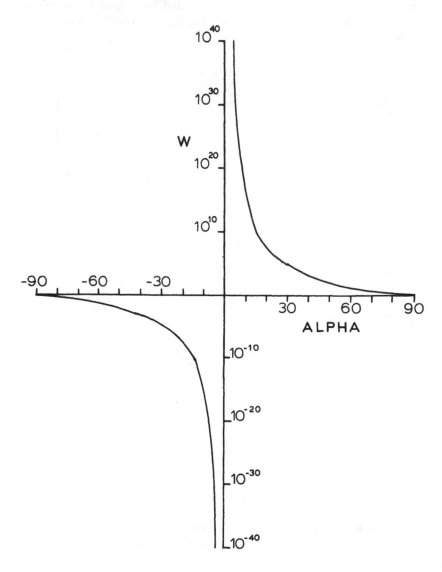

FIGURE 5.4. Relationship between the spiral tangent angle $(\alpha)$ and the whorl expansion rate $(W)$. Positive values of $\alpha$ and $W$ denote convex shell forms, and negative values denote concave shell forms. Note that for flatter and flatter shell forms, $\alpha$ approaches zero, whereas $W$ approaches infinity. *From McGhee 1999d.*

FIGURE 5.5. The dorsal-α–ventral-α theoretical morphospace of planispiral bivalved shell morphology. Along each axis, valve morphologies ranging from convex (positive values of α) to planar (α equal to zero) to concave (negative values of α) can be simulated. All possible permutations of these differential valve morphologies, in combination to produce a hypothetical bivalved shell, are represented here. Note that only those combinations that fall in the shaded region depict biologically feasible brachiopods. *Modified from McGhee 1991.*

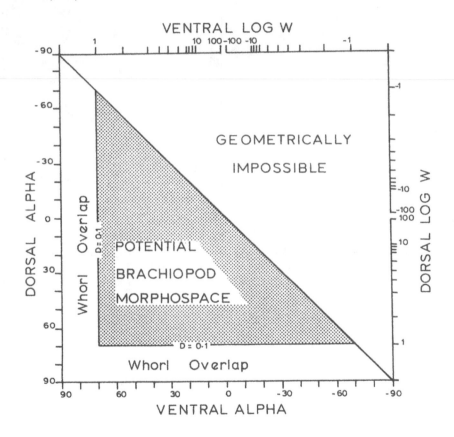

valves (convexi-plane shells) and those with concave ventral valves (convexi-concave shells).

Computer simulations of the distribution of the shell form geometries outlined in figure 5.6 are given in figure 5.7. Note that the hypothetical biconvex brachiopod shell forms given in figure 5.1 are depicted in the lower left corner of the new computer simulations. But the simulations now show shell forms not seen previously, such as those in the upper left corner and the lower right corner.

Examine now the region of plano-convex shell forms (where dorsal α is equal to zero) in the theoretical morphospace of figures 5.6 and 5.7. A different geometric solution to the problem of planar valves, and hence a different theoretical morphospace of planispiral plano-convex shell forms, was created by Ackerly (1992b) in an analysis of shell form

**FIGURE 5.6.** The spectrum of shell form convexities in the theoretical morphospace given in figure 5.5, using conventional brachiopod descriptive morphologic terminology. The inset at the upper right is a stylized representation of figure 5.5, the shaded portion of which (the potential brachiopod morphospace) is expanded in the lower left of the figure to show the distribution of distinctive brachiopod shell morphotypes in the morphospace. *From McGhee 1999d.*

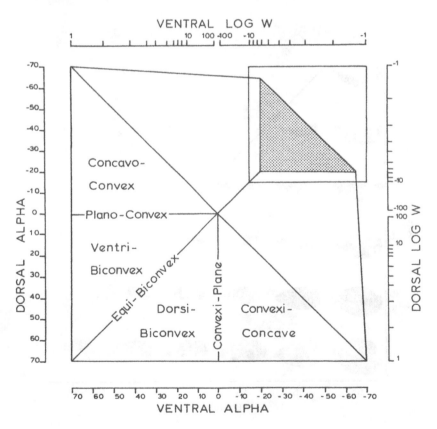

in the bivalve mollusc *Pecten*. Ackerly modeled growth in *Pecten* by using a nearly circular ellipse for a generating curve representing the margin of the shell. This ellipse then was allowed to expand and migrate away from the apex of the shell. The planar, or flat, valve form in *Pecten* may thus be produced with three parameters: (1) aspect ratio, which is a measure of the shape of the elliptical margin, (2) aperture position, which is the relative distance from the valve apex to the ellipse center, and (3) the aperture migration rate. For the convex valve in *Pecten*, the additional parameter of the "radial expansion rate" (Ackerly 1992b:849) was required to measure the curvature of the spiral geometry of the valve (as the apex of the valve is now also a coiling

**FIGURE 5.7.** The computer-produced potential spectrum of both biconvex and nonbiconvex hypothetical brachiopod shell forms in the theoretical morphospace given in figure 5.6. *Modified from McGhee 1991.*

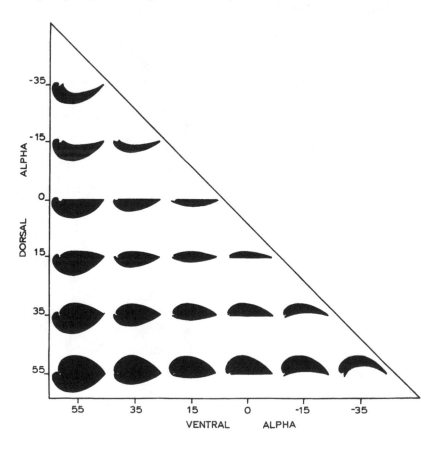

axis). This parameter is in fact the specific growth-rate constant of the spiral (see equation 8.3 in this book and its discussion in chapter 8) and thus is simply a different form of the geometric parameters $W$ and $\alpha$. In the case of allometric growth, the specific growth rate is not a constant but, rather, a variable, as it changes with time (allometric growth models also are discussed in more detail in chapters 8 and 9). Ackerly introduced a fifth and last parameter, allometric trend in radial expansion rate, to simulate allometric growth.

Because Ackerly's (1992b) "aspect ratio" is defined as the ratio between the major and minor axes of the ellipse, this parameter is in essence the same as Raup's $S$, or shape of the generating curve (see chapter 4). Although an important geometric parameter, Ackerly in fact discovered that the aspect ratio is very close to one (i.e., ellipses very close to circular) in a series of empirical measurements of valves of *Pecten* shells, and so this variable may essentially be treated as a constant when simulating shell form in *Pecten*.

Figure 5.8 shows Ackerly's (1992b) theoretical morphospace for planispiral plano-convex shell morphologies, produced by variation in the four parameters, aperture position and aperture migration rate for the planar valve (shown in planar view) and radial expansion rate and allometric trend in radial expansion rate for the convex valve (shown in cross-sectional view). A major difference between this morphospace (figure 5.8) and the plano-convex region of the morphospace given earlier (horizontal region in figure 5.7 where dorsal $\alpha$ is equal to zero) is the "allometric trend in radial expansion rate" axis for the convex valve. This dimension allows ontogenetic changes in the curvature of the valve to be depicted, whereas all the curvatures of the convex (ventral) valves given in figure 5.7 are isometric.

All the theoretical morphospaces of bivalved shell from considered thus far were designed for planispiral shells. The term "planispiral" is a descriptive term understood by all morphologists interested in shell form. From a geometric standpoint, however, the term is actually incorrect. In a typical planispiral ammonite, brachiopod, or scallop shell, only the central cross-sectional, or longitudinal, spiral lies in a plane throughout growth (the $x$-$y$ plane, in the notation of chapter 4). Consider a *Pecten* or scallop shell: the shape of the shell is bilaterally symmetrical about a line drawn from the dorsal apex of the shell to the posteriormost point of the margin. That is, the two halves of the shell are mirror images of each other about this line. Only the spiral trace of this line is a true planispiral; all other spirals in the shell are in fact helicospirals, coiling in $x$-$y$-$z$ space. Thus from a geometric perspective, a "planispiral" shell contains only one planispiral, and all other

**125**

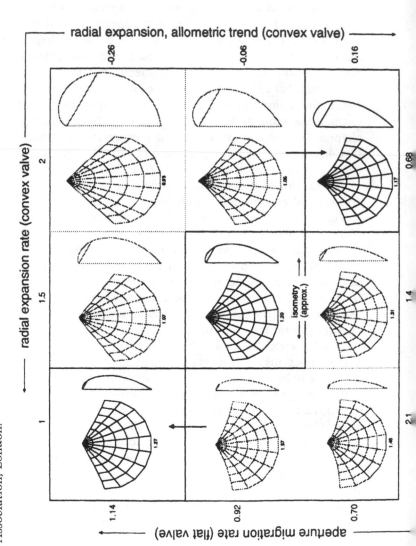

FIGURE 5.8. Ackerly's theoretical morphospace (1992b) of *Pecten* shell forms (planispiral plano-convex shell morphologies), produced by variation in the four parameters "aperture position" and "aperture migration rate" for the planar valve (shown in planar view), and "radial expansion rate" and "allometric trend in radial expansion rate" for the convex valve (shown in cross-sectional view). Only those figures in cells drawn with heavy solid lines (the diagonal of cells from the upper left to the lower right) actually exist in species of *Pecten*; those figures drawn with stippled lines in cells off this diagonal represent hypothetical geometries not found in nature. *From Ackerly 1992b. Reprinted with permission of the Palaeontological Association, London.*

spirals in the shell are helicospirals distributed as mirror images of one another about the central planispiral.

Planispiral shells are produced when the center of the generating curve remains in the x-y plane as the generating curve revolves about the z-axis (the coiling axis); helicospiral shells are produced when the center of the generating curve translates along the z-axis (see chapter 4). In the case of bivalved shells, such helicospiral shells have no spiral section about which the two halves of the shell are mirror images. Helicospiral bivalved shell forms are the most common type of shell morphology found in the group of molluscs called, confusingly, the "bivalves." Bivalve molluscs are a very diverse group in geologic time and are a dominant (and commercially important) part of the shellfish of the modern marine world.

A very interesting theoretical morphospace of helicospiral bivalved shell form was created by Savazzi (1987) using a computer program he published in 1985, which is given in figure 5.9. The theoretical morphospace shown in figure 5.9 has two explicit dimensional parameters: $\Phi$, the angle between the shell aperture (the generating curve) and the coiling axis (the z-axis), and $T$, the translation rate along the coiling axis. An additional parameter, $Y_0$, measures the amount of interumbonal space present in the shell and, though not explicitly shown in figure 5.9, can be visualized as a dimension perpendicular to the two shown.

The parameter $T$ is the same as Raup's (1966). The parameter $Y_0$ in geometric essence is the same as the parameter $B$ discussed earlier with regard to planispiral bivalved shells. In Savazzi 's (1987) model, the amount of space between the umbo of the valve and the dorsal margin of the valve aperture is determined by specifying a certain set of initial coordinates for the generating curve along the y-axis, hence the term $Y_0$, which in turn determines the angular displacement between the two vectors $\mathbf{v_1}$ and $\mathbf{v_0}$. In all the simulations given in figure 5.9, the left and right valves of the hypothetical bivalve mollusc shells are mirror images of each other. Therefore $Y_0$ can be visualized as a single dimension perpendicular to the two shown in figure 5. 9. If the valves of the shells were not mirror images, that is, with different spacings between the umbo and aperture margin in the left versus the right valve, then two dimensions (rather than one) would be required, similar to the dorsal-$B$-ventral-$B$ morphospace considered earlier (figure 5.3).

The interesting parameter $\Phi$ measures a rotation of the plane containing the generating curve, a rotation that is different from the one we considered earlier concerning the difference between a geometric

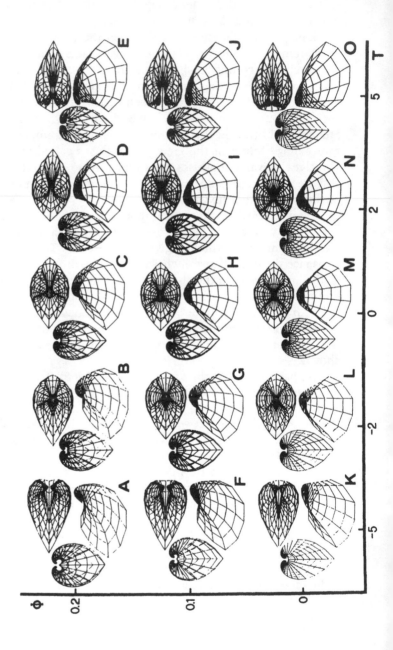

FIGURE 5.9. Savazzi's Φ-T theoretical morphospace of helicospiral bivalved shell morphology (1987). Computer-produced hypothetical shell forms (given in three different views for each simulation) illustrate the morphologic effects of varying the values of the two geometric parameters Φ, the angle between the shell aperture and the coiling axis, and T, the translation rate along the coiling axis. A third parameter, $Y_0$, was also varied to keep interumbonal spacing roughly constant between the left and right valves. *Reprinted from Geometric and functional constraints on bivalve shell morphology by E. Savazzi, from Lethaia 1987, volume 20, pages 293–306, by permission of Scandinavian University Press. Artwork courtesy of E. Savazzi.*

generating curve and a biological generating curve (and the parameter
$B$). Consider once again the circular generating curve, centered on the
$x$-axis and displaced at some distance from the $z$-axis intercept, given
in figure 4.1. Now rotate the generating curve about the $x$-axis itself.
The margins of the generating curve once again lift out of the $x$-$z$ plane
into the $y$-dimension—but not the innermost and outermost marginal
points of the generating curve from the $z$-axis. They remain in the
same position along the $x$-axis. The parameter $\Phi$ measures the angular
amount of the rotation of the generating curve about the $x$-axis. This
parameter in bivalves is thus the geometric equivalent of the gastropod
inclination angle of Linsley (1977) and elevation angle (of the coiling
axis) of Vermeij (1971), discussed in chapter 4. (A brief aside: for those
wishing to use Savazzi's [1985] computer program, note that he uses a
different notation for the parameters than that used in Savazzi [1987].
For example, the parameter $\Phi$ in Savazzi [1987], the Greek letter PHI,
is designated as the computer variable THETA in Savazzi [1985].)

The effect of variation in the parameter $\Phi$ can be seen in the computer simulations in figure 5.9. For each simulation in the morphospace (indicated by A, B, C, and so on, in figure 5.9), Savazzi (1987)
illustrates his hypothetical bivalve molluscs from three aspects: an anterior view (on the left), a dorsal view (at the top), and a lateral view
of the interior of a valve (at the bottom). In terms of spatial geometry,
the anterior view can be considered a perspective of the $x$-$y$ plane
($z$-axis perpendicular to the page), the dorsal view a perspective of the
$y$-$z$ plane, and the lateral view a perspective of the $x$-$z$ plane.

Consider the morphology of the computer simulation "M," in
which both $T$ and $\Phi$ are zero (figure 5.9). This particular simulation is
almost a planispiral bivalved shell, where the two halves of the shell
are mirror images about a line drawn from the umbo to the posterior
margin. It is not planispiral only because the initial generating curve
was chosen to be asymmetrical about the $x$-axis, which is the common
condition among bivalve molluscs. Note the change in this initial
morphology, particularly in the umbonal region of the shells, as values
of the parameter $\Phi$ are increased from simulation "M" to "H" to "C."
Without any change in the translation rate ($T$ remains zero), the anterior regions of the shell decrease in relative size while the posterior
regions increase, and the umbos of the two valves twist toward each
other with greater and greater angles. The parameter $\Phi$ is thus particularly useful in modeling prosogyral and opisthogyral shell forms, common in the bivalve molluscs.

The effect of changes in the values of the parameter $T$ can be seen
in the progressively more helicospiral shell forms proceeding from

"M" to "N" to "O" and from "M" to "L" to "K." Note that the geometric effect is the same in the positive $T$ direction and the negative $T$ direction, that is, that shell form "K" is simply the reverse of "O" (figure 5.9). In a real bivalve mollusc, the anterior of the animal is quite different from the posterior. In terms of geometric simulations of shell form, however, the anterior and posterior ends of the shell are arbitrary designations, thus only half the theoretical morphospace need actually be considered in order to include all possible morphologies that might be found in actual bivalve molluscs (Savazzi 1987).

In conclusion, a variety of different theoretical morphospaces have been created to simulate the morphologies commonly found in organisms having two valves articulated together to form a bivalved shell. The shell forms discussed up to this point are hypothetical, geometric abstractions. Let us now examine what nature has actually produced in the evolution of bivalved shell forms.

## THE EXPLORATION OF PLANISPIRAL BIVALVED MORPHOSPACE

Let us return to the theoretical continuum of planispiral biconvex bivalved geometries given in figure 5.1. All the morphologies shown in figure 5.1 are hypothetical, computer produced, and represent a sample of the theoretical morphologies potentially available to organisms that produce planispirally coiled valves articulated together in a single bivalved shell, such as the biconvex Brachiopoda. Not surprisingly, as it was designed for that purpose, this theoretical morphospace has been extensively used to examine the evolution of shell form in the Brachiopoda by McGhee (1978b; 1979a,b; 1980a,b,c; 1982b; 1991; 1995a,b; 1999a,b,c,d).

Actual measurements of a sample of 324 biconvex brachiopod species (McGhee 1980b), each representing a different genus, reveal the actual utilization pattern of this group of organisms in the geometric continuum of hypothetical forms illustrated in figure 5.1. The frequency distribution of potential geometries actually used by the biconvex Brachiopoda is given in figure 5.10. The peak, or most frequently used, shell form in the biconvex brachiopods has a dorsal valve convexity of $W$ equal to $10^5$ and a ventral valve convexity of $W$ equal to $10^3$ and is thus inequivalved (displaced upward from the diagonal line shown in figure 5.10, which depicts the region of equivalved shells in the geometric continuum). The frequency distribution of morphologies is positively skewed, with the majority of brachiopod shells occurring in regions of low $W$ magnitudes but with the distribu-

tion tailing outward to the high-magnitude regions of $W$ values ranging from $10^{10}$ to $10^{12}$. Conversely, no biconvex brachiopods are found
with very low magnitudes of $W$, such as $10^1$ to $10^2$.

Figure 5.10 gives the actual frequency distribution of biconvex
brachiopod shell morphologies produced by nature in the evolution
of the group. One may then ask, what, if anything, is the functional
significance of this frequency distribution? Actual biconvex brachiopods occupy only a limited region in the theoretical continuum of potential shell geometries; that is, there are conceivable geometries that

**FIGURE 5.10.** The frequency distribution of actual shell morphologies
found in 324 genera of biconvex brachiopods, in the theoretical morphospace given in figure 5.1. Taxa that fall along the diagonal line in the morphospace are equibiconvex. Contours measure the increase in density of
genera per unit area of the plot. *From McGhee 1980b.*

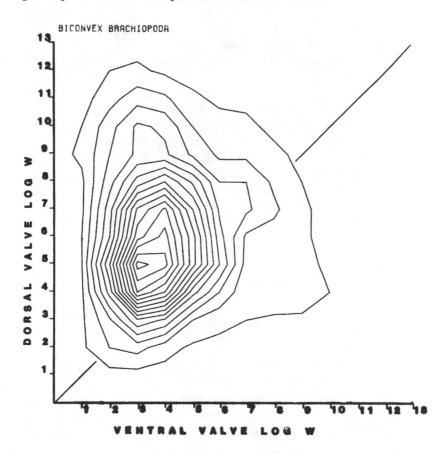

cannot be found among the Brachiopoda in nature (cf. figures 5.1 and 5.10). It could be argued that this is due to chance effects alone, that the brachiopods simply have not had sufficient enough time in their evolutionary history to evolve the potential morphologies not now present in the group. Alternatively, there might be specific functional and mechanical reasons behind the current range and frequency of morphologies found among biconvex brachiopods.

Rather than chance, I have argued that the observed distribution of brachiopod morphologies is directly related to specific biological limitations imposed by the geometry of shell growth, function, and evolutionary history of the group (McGhee 1980b,c; 1999b,c). Biconvex brachiopods tend to produce shells that have maximum internal volumes to minimum shell surface areas; that is, they approach a spherical form as closely as possible within specific geometric constraints. The boundaries of these constraints, which define the biconvex brachiopod morphospace, are given in figure 5.11. The great majority of biconvex brachiopod shells cluster in the lower left corner of the morphospace, in regions of low magnitudes of $W$—which is the region of maximum internal volume to minimum surface area in the geometric space. The brachiopods are prevented from even further reductions in $W$ magnitudes (even closer approximations to spherical forms) by limitations imposed by their system of articulating the two planispirally coiled valves into a single functional bivalved shell. These articulation limitations involve minimum space constraints in the ventral umbonal region of the shell, which limit the development of shells that are strongly dorsi-biconvex to convexi-plane (minimum departure from equiconvexity indices shown in figure 5.11), and the fact that whorl overlap must be absent in both valves if they are to be articulated together into a single functioning shell (maximum distance from coiling axis indices in figure 5.11; for an illustration of hypothetical valve form in this region, see figure 5.1). The converse of this argument is that there may be shell geometries with such large surface areas and small internal volumes that those geometries may be bioeconomically unattainable for biconvex brachiopods. Hence the last set of limitations (area-volume ratio indices in figure 5.11) seen in the high $W$ magnitude region of the biconvex brachiopod morphospace (McGhee 1980b, 1999b).

Consider now the additional dimensions of dorsal $B$ and ventral $B$ in the theoretical morphospace (figure 5.3). The actual frequency distribution of biconvex brachiopod shell forms in this region of the theoretical morphospace is given in figure 5.12, as measured by the same sample of 324 species given in figure 5.10. The three cells corresponding to the zero-*B* row for the dorsal valve contain the majority of the sample. In contrast, sample sizes for the dorsal-negative-*B* and ventral-negative-*B* cell, and the dorsal-negative-*B* and ventral-zero-*B* cell were too small to register a measurable percentage, and these two cells (see figure 5.3) are empty in figure 5.12. For the entire sample, the dorsal-zero-*B* and ventral-negative-*B* cell contains the greatest percentage of biconvex brachiopods (figure 5.12). The second largest percentage lies in the three cells defined by the ventral-positive-*B* row.

What is the significance of these distributions? Biconvex brachiopods may be subdivided into two major groups on the basis of their hinge mechanism: either strophic or nonstrophic. Strophic brachiopods possess a true hinge line; that is, the hinge axis (the axis about

**FIGURE 5.11.** The biconvex articulate brachiopod morphospace. Boundaries of the space are discussed in the text. *From McGhee 1980b.*

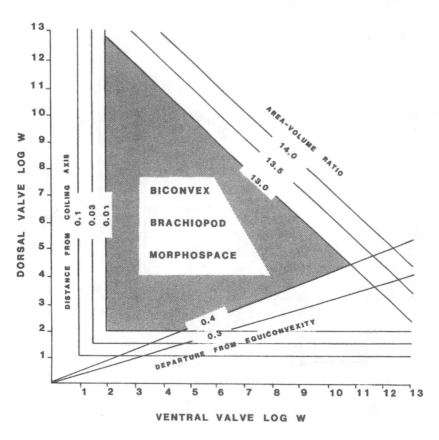

which the articulated valves rotate in opening and closing the shell) corresponds to a straight line segment of the growing posterior margin of the valves, a segment that forms the actual lever fulcrum for the valves (see figure 5.13A). In contrast, nonstrophic brachiopods do not possess a hinge line, in that no segment of the growing posterior margin of the valves corresponds to the hinge axis about which the valves rotate (figure 5.13B).

In the spectrum of forms shown in figure 5.3, nonstrophic shells usually have morphologies that fall in the zero $B$ range for the dorsal valve, and strophic shells more commonly have dorsal valves with zero or positive values of $B$. Thus the two observed morphologic distri-

**FIGURE 5.12.** The frequencies of the shell forms, in terms of the spectrum of possible $B$ values, which are actually found in biconvex brachiopods. Compare with figure 5.3 for an illustration of the shell forms found in each cell. *From McGhee 1980b.*

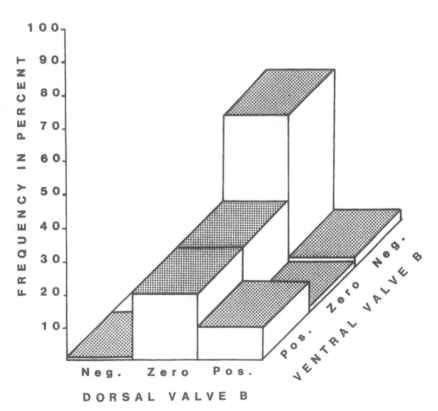

**FIGURE 5.13.** Illustration of a strophic brachiopod (figure 5.13A) and a nonstrophic brachiopod (figure 5.13B). Growth of the posterior margin of the strophic brachiopod takes place coincident with the hinge axis of the shell (horizontal line). In the nonstrophic brachiopod, no segment of the growing posterior margin corresponds to the line of the hinge axis of the shell.

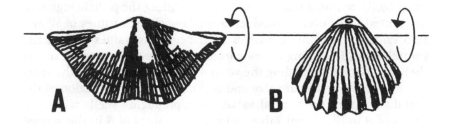

butions (dorsal-zero-$B$ and ventral-negative-$B$ cell containing the largest percentage, the three cells defined by the ventral-positive-$B$ row the second largest) generally coincide with typical nonstrophic and strophic shell morphologies, respectively.

Now we can ask, what is the functional significance, if any, of these differences in shell morphology? The effect of a positive $B$ value in either valve is to increase the internal volume of the shell, particularly in the umbonal region. The effect of a negative $B$ value is exactly the reverse. The rarity of brachiopod shells with negative $B$ values for both the dorsal and ventral valves is a function of the limitations imposed by their articulation strategy, as sufficient space must be present in the ventral umbonal region to allow the insertion of the dorsal umbo (for dorsal-zero-$B$ shells) or the deeper insertion of the already inserted dorsal umbo (for dorsal-negative-$B$ shells) during the process of opening the valves (McGhee 1980b,c).

On the one hand, the great abundance of the dorsal-zero-$B$ and ventral-negative-$B$ type of shell among nonstrophic brachiopods and the large positive $B$ values seen in the ventral valves of strophic brachiopods are primarily related to problems of maintaining stability in an epibenthic mode of life (McGhee 1980c, 1999c). For strophic brachiopods, a strong positive correlation exists between convexity (as measured by $W$) and height of the interarea (as measured by $B$) in the ventral valve of the shell; that is, the flatter the convexity of the ventral valve (high $W$ magnitude) is, the larger the ventral valve interarea (high positive $B$ magnitude; see the discussion in McGhee 1980c) will be. A large positive $B$ in the ventral valve, with the resultant high ventral valve interarea (see figure 5.3 for an illustration), in combination with the straight hinge line of the strophic shell, produces a triangular surface that was used as a resting base by many strophic brachiopods (McGhee 1980c, 1999c).

Nonstrophic brachiopods, on the other hand, either attach themselves to the substrate by means of a pedicle or lie free on the substrate. Those with a functioning pedicle can easily attach to the substrate by flattening the dorsal valve while coiling the pedicle region of the ventral valve over it, producing shells with high values of $W$ and zero values of $B$ in the dorsal valve and negative values of $B$ in the ventral valve. Free-lying forms commonly weight the anterior of the shell, coiling and expanding the ventral umbonal region over the anterior margin of the dorsal valve and tucking the umbonal region of the dorsal valve into the ventral valve, again producing shells with zero values of $B$ in the dorsal valve and negative values of $B$ in the ventral (McGhee 1980c, 1999c).

Finally, let us consider the enigma of the nonbiconvex articulate brachiopods, which possess shell forms virtually the exact opposite in morphology as those possessed by the biconvex articulate brachiopods. I argued earlier that biconvex brachiopods develop spherical shell forms with maximum internal volumes to minimum shell surface areas, shell form geometries with low area-volume ratios most favorable to their feeding system of water filtration via a lophophore enclosed in the shell. In contrast, nonbiconvex brachiopod shell morphologies, commonly found in the Orthida and Strophomenida, obviously do not maximize internal shell volume to external shell surface area. Exactly the opposite is seen in many of the extremely flattened and compressed concavo-convex morphologies found in the Strophomenida, suggesting that the extinct strophomenides in particular may have had a quite different mode of life than do extant biconvex brachiopods.

Actual measurement data for nonbiconvex brachiopods are currently being collected and are not available to map the distribution of these enigmatic shell forms in the new geometric continuum given in figure 5.6. However, in the spirit of Raup (1966), who mapped out the rough distribution of gastropod shells, bivalve mollusc valves, and brachiopod valves in his original theoretical morphospace of univalved shell form in the absence of any measurement data, we can make a similar rough mapping of nonbiconvex brachiopod shell form. The majority of the Orthida fall in the region delimited by bracketing dorsal $\alpha$ values between 0° to +15° along the $y$-axis and extending these two limits over to the diagonal and down to the same limits along the ventral $\alpha$ axis, in figure 5.14. In conventional brachiopod terminology, the majority of the Orthida occur in the highly ventri-biconvex to plano-convex region of the y-axis in figure 5.6, extending over and down to the highly dorsi-biconvex to convexi-plane region along the x-axis in figure 5.6 (cf. figure 5.7 for an illustration of these shell forms). The Strophomenida are even farther displaced from the lower left corner of the morphospace in figure 5.6 and are found almost exclusively in the plano-convex to concavo-convex corner of the morphospace (dorsal α less than or equal to 0°) or in the convexi-plane to convexi-concave corner of the morphospace (ventral α less than or equal to 0°). More specifically, in the Strophomenida, the productidines preferentially occur in the former region, and the distinctive highly compressed shell forms of the strophomenidines are spread along the upper diagonal margin of the morphospace in figure 5.6 (cf. figure 5.7 for an illustration of these shell forms).

If the techniques of theoretical morphology can contribute to the functional analysis of these enigmatic shell forms, they must await the actual mapping of the frequency distribution of nonbiconvex shell form in the morphospace. At present, however, it is clear that these enigmatic brachiopods were not maximizing internal shell volume relative to shell surface area. Hence they were probably not using the same system of water filtration in the shell, via the lophophore, seen in modern living biconvex brachiopods (see James et al. 1992), and thus the mode of feeding in these extinct nonbiconvex forms may have been very different from that seen in extant brachiopods.

Even though data for nonbiconvex brachiopods are not yet available, Ackerly (1992b) explored the planispiral plano-convex region of theoretical shell morphospace for seven species of the bivalve mollusc *Pecten*. Consider Ackerly's theoretical morphospace for these shell forms given in figure 5.8. In his analysis of measurements of 23 individuals from seven separate species of *Pecten*, Ackerly found that only those morphologies illustrated in the three diagonal cells, located from the upper left to the lower right in the morphospace, actually exist in nature. Hypothetical *Pecten* shells shown in the six remaining cells, located above and below this diagonal, are entirely conceivable and can be produced by computer simulation but nevertheless do not occur in nature. The diagonal of existent shell forms in figure 5.8 is produced by a surprising pattern of covariation in the parameters chosen to define the dimensions of the morphospace. Ackerly found a strong inverse relationship between actual measurements of aperture position and aperture migration rate for planar valves, and radial expansion rate and allometric trend in radial expansion rate for convex valves, in individuals from the seven species of *Pecten* considered. The functional

137

significance of this covariation is not entirely clear, but Ackerly considers it likely to be due to the swimming behavior of scallops. The
narrower and more hemispherical shells in the upper right of the morphospace have shapes that would make it difficult to produce the dor-
sally directed jets of water used by these bivalve molluscs in swimming. The broader and more compressed shells in the lower left of

FIGURE 5.14. The predicted distribution of nonbiconvex shell forms in the
orthide, productidine, and strophomenidine brachiopods in the theoretical
morphospace given in figure 5.6. The open, unpatterned square in the
lower left corner is the region occupied by biconvex brachiopods (see figure
5.10). Question marks indicate the unknown maximum departure of nonbiconvex brachiopods from the minimum shell surface area to maximum
internal volume region of the morphospace (lower left corner).

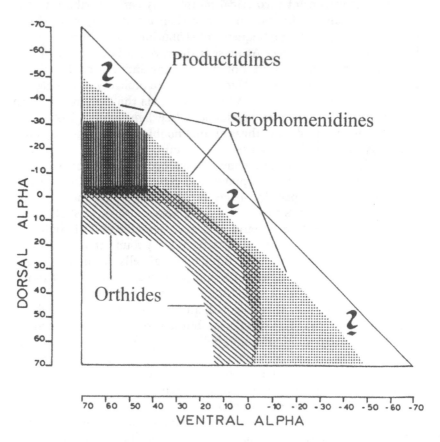

the morphospace may have insufficient internal volumes to produce sufficient water storage capacities needed for propulsion.

Many years ago, Cowen and Rudwick (1967) and Rudwick (1970) suggested that some gently concavo-convex strophomenidine brachiopods might have been able to swim, analogous to modern scallops. These brachiopods possessed extremely compressed shell forms, as discussed earlier (see also figure 5.14), and are most similar to the shell morphology seen in the extreme lower left cell of the *Pecten* morphospace (figure 5.8). Thus, if Ackerly's (1992b) reasoning holds, it is extremely unlikely that any of these brachiopods had the ability to swim, and certainly not like scallops.

## THE EXPLORATION OF HELICOSPIRAL BIVALVED MORPHOSPACE

What can I say? Logically, I should now complement the previous section on the exploration of planispiral bivalved morphospace. Yet I find that I have nothing to write about, no interesting studies of the distribution of actual organic form in theoretical morphospace to discuss. And the problem is not that helicospiral bivalved morphospaces do not exist, as I talked about a very interesting one earlier, given in figure 5.9, and more can be easily imagined.

The problem here is that no one has bothered to explore the distribution of naturally occurring shell form in a helicospiral bivalved morphospace! Savazzi (1987) offers a few predictions about where actual bivalve molluscs might occur in the theoretical morphospace of figure 5.9: the arcoids probably occur in the lower right corner of the morphospace, the venerids in the upper right corner, and so on. Thus the state of helicospiral bivalved morphospace exploration is similar to that of univalved morphospace exploration in 1966, when Raup predicted where typical gastropods and so on might fall in "the Cube" morphospace. He had no actual measurement data at that time, but other researchers were quick to rectify that situation, as pointed out in the exploration of univalved morphospace sections of chapter 4. There was no similar research response for the bivalve molluscs. This is truly surprising, at least to me, because the bivalve molluscs have a long and interesting evolutionary history, and today they constitute the majority of the shellfish in oceans. And the number of paleobiologists and biologists who study the bivalve molluscs is enormous! Thumbing through my copy of the *Directory of Paleontologists of the World* (Doescher 1989), I find that the primary area of research of 988

139

paleontologists is the molluscs. Of those, 312 paleontologists, or 32% of the mollusc workers, focus specifically on fossil and recent bivalves. The number of modern biologists who work with bivalve molluscs must be even larger, particularly because these animals have commercial importance.

Why, then, has no one explored the distribution of the bivalve molluscs in theoretical morphospace? I have no answer to that question. I would simply point out here, for the consideration of young graduate students searching for an interesting dissertation topic, that the analysis of the evolution of the bivalve molluscs in a theoretical morphologic context is wide open—a ripe fruit waiting to be seized!

# VI

# Step by Step: Theoretical Morphospaces of Discrete Growth Systems

Many organisms grow by a process of discrete, episodic, stepwise additions of parts or elements, rather than continuous accretion to an already existent morphology. The tiny single-celled foraminifera, for example, produce shells (called tests) that often have beautiful planispiral or helicospiral forms. However, the planispiral or helicospiral test of a foraminiferid is quite different from the planispiral or helicospiral shell of an ammonite or a gastropod. The test of a foraminiferid is composed of a series of small globes or chambers, discrete units that are added episodically during the growth of the cell. The shell of an ammonite or gastropod is composed of thousands of thin layers of shell material added continuously at the growing edge of the organism's mantle.

Some may quibble with this characterization of growth as "discrete" versus "continuous." They might point out, for example, that a gastropod does not in fact continuously add new material to its shell but does so in a series of growth increments that can often be seen in visible growth lines on the exterior of the shell. I consider the distinction to be a useful one, however, and I think that the concept is clear. A "discrete" growth increment of thin material added at the edge of a large shell is very different from the discrete addition of an entire new globular chamber to a preexisting smaller globular chamber. One is the addition of an accretionary unit, the other is the addition of a new part or element.

And last, some organisms grow in both a continuous and a discrete fashion. Many echinoderms enlarge the plates of their skeletons by continuous accretion at the margins of the plate. Once certain plates have grown to a critical size, however, the echinoderm then discretely adds an entirely new, much smaller plate to its exoskeleton. The new plate then begins to grow in an accretionary fashion, and this mixed pattern of growth is repeated over and over during the life of the echinoderm.

Discrete growth systems are more difficult to model than accretionary growth systems. Perhaps the ultimate nightmare of any modeler of morphogenesis is organisms that molt, whose entire skeleton is episodically shed and regrown, often with the new skeleton possessing parts or elements not previously present in the organism. Such a discrete mode of growth is used by the arthropods, and the arthropods are the dominant form of metazoan life on the planet (in terms of both sheer numbers and species diversity).

## THEORETICAL MORPHOSPACES OF CHAMBERS AND CYLINDERS

Foraminifera commonly produce spiral forms during their growth, but most do so in a discrete growth process of serial chamber additions. Theoretical morphogenetic models of foraminiferal growth have been the subject of studies by Davaud and Wernli (1974), Bayer (1978a), Brasier (1980; 1982a,b; 1984; 1986), De Renzi (1988, 1995), and Signes and colleagues (1993). Of these, Davaud and Wernli (1974), De Renzi (1988, 1995), and Signes and colleagues (1993) produced computer simulations of foraminifera form that can be modified to form theoretical morphospaces of hypothetical test morphology; indeed, De Renzi (1995) and Signes and colleagues (1993) are well on the way in this direction.

The first explicit theoretical morphospace of foraminiferal form was created almost three decades ago by Berger (1969). Although the model is planispiral, Berger was also able to "simulate" three-dimensional "trochospiral" (helicospiral) morphologies as well, by altering the aspect view of the simulation. The model also assumes spherical chambers, as planispiral and trochospiral tests composed of spherical chambers are particularly prevalent in planktic foraminifera. The theoretical morphospace has three parameters:

142

1. *a-angle* = the "angle of advance," defined as the angle between the two lines connecting the x-y origin with two successive chamber centers.
2. *o-lap* = the "amount of overlap" between two successive
   chambers, defined as the ratio of the radius of a chamber to
   the distance between its center and the center of the succeeding chamber.
3. *q-ratio* = the ratio between successive chamber radii.

Construction of a hypothetical foraminiferal test, or shell, using
these three parameters is illustrated in figure 6.1. Start with a single
spherical chamber (the "proloculus," in foraminiferal terms) centered
on the x-axis with its radius, $R_1$, set as the distance from the x-y origin
(figure 6.1A). The next chamber position is determined by the parameter *a-angle*, the angle between a line from the origin through the center of the first chamber and a line from the origin through the center
of the second chamber (figure 6.1B). The distance from the center of
the first chamber to the center of the second chamber is taken as the
radius of the first chamber ($R_1$) divided by the parameter *o-lap*, the
amount of overlap (figure 6.1C). The radius (hence the size) of the second chamber ($R_2$) is taken as the radius of the first chamber ($R_1$)
multiplied by the parameter *q-ratio* (figure 6.1D). This procedure is
repeated to determine the position of the third chamber with respect
to the origin, the distance between the centers of the second and third
chambers, and the size (radius $R_3$) of the third chamber (figure 6.1E).
Repeat the procedure to add a fourth chamber (figure 6.1F).

The successive circles shown in figure 6.1F represent two-
dimensional sections, lying in the x-y plane, of a succession of three-
dimensional spheres. The simulations may be made more realistic by
eliminating the arcs of the circles in other circles, as new chamber
construction only takes place outside the chamber walls of the previously existent chamber in foraminifera. Furthermore, the planispiral
simulations may be made to appear "trochospiral" (gently helicospiral) and thus more realistic by specifying a "dorsal" surface, in
which the arc of the initial chamber fully interpenetrates the globe of
the second chamber, the second chamber that of the third, and so on
(figure 6.1G), versus a "ventral" view, in which the boundaries between the successive chambers are linear and do not interpenetrate
(figure 6.1H).

The growth parameters *a-angle, o-lap*, and *q-ratio* are held constant
during successive chamber formations; thus the growth model is isometric. As the *a-angle* in particular is a constant, the spirals produced

FIGURE 6.1. Eight steps (A through H) in the creation of a hypothetical "trochospiral" (heliospiral) planktic foraminiferal test. See the text for discussion. *From Berger 1969. Used with permission of SEPM—Society for Sedimentary Geology (formerly Society of Economic Paleontology and Mineralogy).*

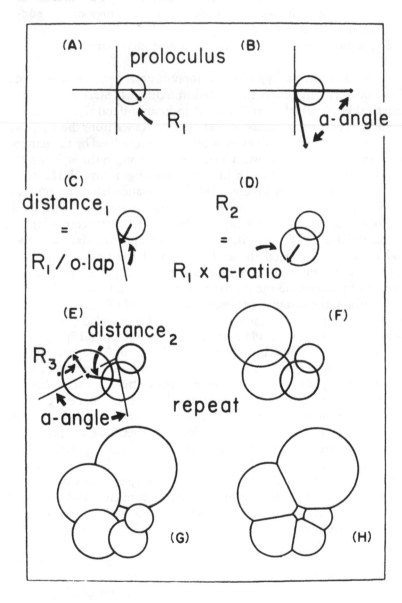

are logarithmic (see chapter 8 for a discussion of logarithmic, or equiangular, spirals). The logarithmic spirals of foraminifera are composed of discrete chamber units, however, and so are different from the logarithmic spirals generated by the continuous differential equations used to simulate shell form in the snails or ammonites discussed in chapter 4.

The discrete nature of growth in planktic foraminifera is reflected in the values chosen for the three parameters designed by Berger (1969). The parameter *a-angle* is an angle measure, and angles are measured as continuous rational numbers, such as degrees divisible into minutes, and minutes divisible into seconds, and so on. The parameter *a-angle* is in fact measured in terms of integers, and not rational numbers, in Berger's model. In the computer simulations, the values of *a-angle* were chosen to produce hypothetical foraminiferal tests with 3, 4, 5, and 6 chambers per whorl; thus the values of *a-angle* were always 120°, 90°, 72°, or 60° (corresponding to 3, 4, 5, and 6 chambers per whorl, respectively), and nothing between those values.

Likewise, the values of the parameter *q-ratio* are not measured in an arithmetic fashion, but as a geometric sequence of the numbers 1.1, 1.21, 1.331,; that is, as powers of the base 1.1:

$$q\text{-}ratio = (1.1)^1, (1.1)^2, (1.1)^3, \cdots , (1.1)^n \qquad (6.1)$$

The values of the parameter *o-lap* are counted in a similar fashion, but the geometric sequence is 0.4 times $(1.1)^n$ rather than just $(1.1)^n$, as in the parameter *q-ratio*. Berger (1969:1371) justified the usage of a geometric series for values of *q-ratio* and *o-lap* "because arithmetic steps led to a very slow change in the morphological aspects of the models." Because the powers in the series are integers, these two parameters are measured in a "discrete" integer system as well.

Two sections from the three-dimensional theoretical morphospace of Berger (1969) are given in figure 6.2. The upper diagram illustrates the range of hypothetical foraminiferal morphologies present in the morphospace where the value of *a-angle* is set to produce four chambers per whorl and the values of *o-lap* and *q-ratio* are allowed to vary. Note that the last-formed chamber almost totally encloses all previous chambers in the lower right corner of the diagram. In the opposite corner (upper left) there are no simulations at all, and this region is labeled a "forbidden" zone (figure 6.2). It is "forbidden" because the values of the parameters *o-lap* and *q-ratio* in this region produce successive chambers that do not touch one another in space, isolated

FIGURE 6.2. Two selected two-dimensional sections of Berger's three-dimensional theoretical morphospace of foraminiferal test morphology (1969). The "forbidden range" regions of the morphospace indicate parameter combinations that produce successive chambers that do not touch one another. The diagram at the top illustrates hypothetical planispiral foraminiferal tests having four chambers per whorl. The diagram at the bottom illustrates hypothetical trochospiral foraminiferal tests having six

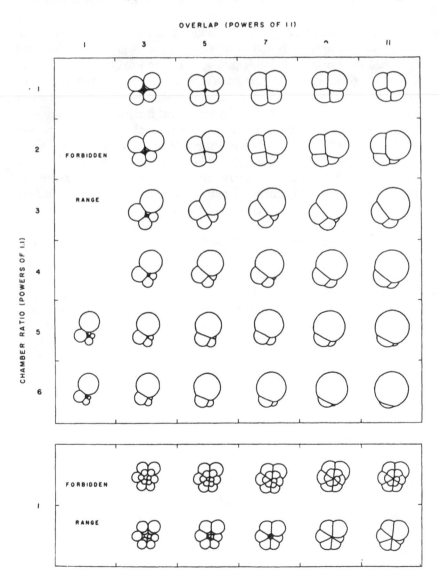

chambers per whorl in two views: dorsal aspect in the upper row and ventral aspect in the lower row. Values of the two parameters "overlap" and "chamber ratio" are measured in powers of 1.1 (see the text for discussion). *From Berger 1969. Used with permission of SEPM—Society for Sedimentary Geology (formerly Society of Economic Paleontology and Mineralogy).*

spheres that cannot be cemented together to form a foraminiferal test. Thus Berger's theoretical morphospace is able not only to simulate existent and nonexistent foraminiferal form but also to have parameter combinations that are geometrically impossible, like the "inside-out" brachiopod region in McGhee's (1991) theoretical morphospace, discussed in chapter 5 (see figure 5.5).

Hypothetical foraminiferal tests with six chambers per whorl are given in the bottom diagram in figure 6.2. Here "trochospiral" tests are simulated by showing the margins of all the chambers formed along the trace of the spiral, from the smallest at the top of the spire to the largest at the basal margin, in a "dorsal" view and having the margins of the last-formed whorl progressively eclipse those of previous whorls in a "ventral" view (figure 6.2).

Berger (1969) did not take the next step in theoretical morphologic analysis, that of plotting the positions of actual foraminifera in the theoretical morphospace of hypothetical forms. He did explore some of the geometric properties of the morphospace's various regions and compared theoretical growth curves of test size versus age with actual measurements taken from living foraminifera, which appear to follow either linear or decreasing growth curves. Berger also found that even though all three parameters are needed to define form, only the *q-ratio* parameter is necessary to characterize growth relationships between test surface area and internal volume and in the thickness of the walls of the chambers. Although he states that the *q-ratio* parameter is "easily measured as the ratio of the diameter of the test with n chambers to the diameter with n-1 chambers" (Berger 1969:1383), he does not give any measurements in the study.

Berger's (1969) model simulates isometric growth in foraminifera. In a lengthy paper that is otherwise morphometric, Scott (1974:140–143) devoted a small section to the theoretical morphologic modeling of anisometric growth in planktic foraminifera. Anisometric growth in foraminifer was also examined by Bayer (1978a) and De Renzi (1988, 1995).

Berger's (1969) model also is confined to the simulation of morphologies found in planktic foraminifera, namely, planispiral to helicospiral skeletons composed of spherical chambers. Brasier (1980, 1982a) conducted a comprehensive overview of all foraminiferal skeletal morphology, not just those of the planktic forms. Stating that "the tests of foraminifera exceed in architectural variety those of any other invertebrate group" (Brasier 1982a:1), he carefully classified the various growth modes that produce not only spirally arranged chambers that touch or overlap slightly but also those in which the chambers successively completely overlap all previous chambers, those that add chambers in linear rows with no spiral component, and those that grow in a nonseptate mode more similar to that seen in accretionary growth systems.

In his early work, Brasier (1980) produced a generalized working-model theoretical morphospace of foraminiferal form using four dimensional parameters: (1) "rate of chamber expansion," (2) "degree of chamber compression and overlap," (3) "rate of growth translation," and (4) degree of "extension of growth along coiling axis." A schematic sketch of the various types of foraminiferal skeletal forms that can be produced by varying these four parameters in a qualitative fashion is given in figure 6.3. In addition to Berger's (1969) planispiral and trochospiral form illustrated in figure 6.2, we now see the extended range of serially arranged chamber forms (uniserial, biserial, triserial) that have no coiling component of growth, as well as those form that do coil but do so involutely (up to the point where the test is completely involute and only the last formed chamber is visible), and those forms that "translate" along the coiling axis in *both* directions (and not just one, as in the case of helicospirals), producing the "fusiform" type of foraminiferal test (figure 6.3).

Brasier does not further develop and quantify the theoretical morphospace given in figure 6.3 in his later work but, rather, turns to an analysis of the geometric properties of foraminiferal tests, particularly the two internal geometric variables: "minimum line of communication from the back of the proloculus to the nearest aperture," termed the *MinLOC*, and the "maximum line of communication from the most remote point of a distal chamber to its nearest aperture," termed the *MaxLOC*. Stating that "the MinLOC method provides another way of looking at foraminifera, from the inside" (Brasier 1982b:103), he then conducted a series of studies of evolutionary trends in the reduction of internal lines of communication in the test (reductions in *MinLOC* and *MaxLOC*) in foraminifera, and the functional significance of these evolutionary trends (Brasier 1982a,b; 1984, 1986).

The most comprehensive return to the external aspects of foraminiferal form generation since Berger (1969) was that of Signes and col-

**FIGURE 6.3.** Brasier's generalized theoretical morphospace of foraminifera (1980), in which foraminiferal form is determined by four qualitative parameters of growth (indicated by arrows in the figure; see the text for discussion). All the morphologies shown are hypothetical axial sections of foraminiferal tests, with the exception of the four equatorial sections given in the two boxes at the top of the figure. Traditional morphologic terminology used to describe different types of foraminifera is also given in the figure. *From Brasier 1980. Copyright © 1980 by Chapman and Hall; used with permission.*

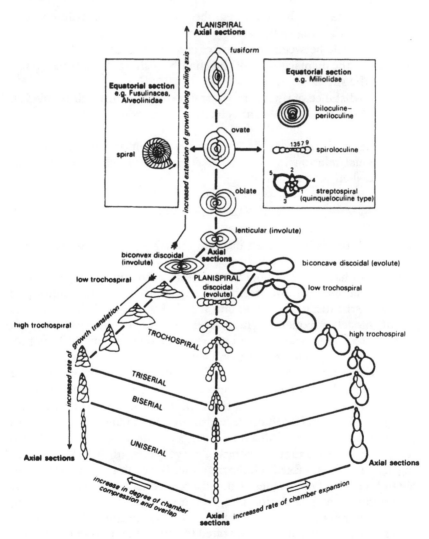

leagues (1993), who designed a theoretical morphogenetic model by making two basic assumptions concerning foraminiferal growth: (1) the shape of the chambers in the shell remains constant with growth, and (2) the volume of each new chamber increases in a constant proportion to the preexisting volume of the shell. Their model has four parameters:

1. $Kt$ = the ratio between chamber volume and the volume of the preexisting shell.
2. $\phi$ = angle between consecutive chambers in the shell.
3. $Ky$ = the displacement of the chambers along the coiling axis of the shell.
4. $D$ = distance of the chamber center to the coiling axis divided by the radius of the same chamber.

The heart of the theoretical morphogenetic model is the constant proportional relationship ($Kt$) between successive chamber volumes ($CV$) and shell volumes ($SV$):

$$(CV)_{n+1} = (Kt)(SV)_n \tag{6.2}$$

where $n$ is the index of each discrete growth step. Because shell volume is a function of the volumes of the chambers that comprise it, the rate of change of chamber volume with each growth step is a function of chamber volume itself, and exponential growth results (see chapter 8, the discussion of equation 8.2). The two basic assumptions of the model (exponential increase in chamber volume with each growth step, and constancy of chamber shape with growth) thus produce a growth system that is isometric.

The other parameters in the model relate to the positioning of the chambers in three-dimensional space. The parameter $\phi$, the angle between consecutive chambers in the shell, is the same as the parameter *a-angle* in Berger's (1969) model. Although the Signes and colleagues (1993) model is one of discrete chamber additions, it is similar to Raup's (1966) continuous generating-curve model in that it has a coiling axis and thus a fixed-reference frame. For example, the parameter $Ky$ in Signes and colleagues is the discrete equation equivalent of the parameter $T$, the translation rate, in Raup's logarithmic spiral model, and the parameter $D$, though not the same, is similar in both models.

Signes and colleagues (1993) stated that the chief goal of their study was to use the model to examine the functional properties of the different hypothetical morphologies produced by varying the parameter

150

values of the model, and they did examine the various relationships of shell surface area, shell internal volume, and size effects in both, with differing shell structural types. They did not explicitly figure a four-dimensional morphospace, using the model parameters $Kt$, $\phi$, $Ky$, and $D$ as dimensional axes, but they did talk about the construction of such a theoretical morphospace (Signes et al. 1993:77–79) and even discussed the existence of four nonoverlapping regions in such a theoretical morphospace.

First, Berger's (1969) "forbidden region," where the chambers of the "shell" are isolated spheres that do not touch one another, is stated to occur in the morphospace where the ratio of the radius of a chamber to the distance from its midpoint and the midpoints of the preceding or succeeding chambers is less than $(Kr + 1)^{-1}$, where $Kr$ is defined as the ratio between the radii of consecutive chambers and, in the model, is equal to the cube root of $(Kt + 1)$.

Second, the region of shells with completely involute chambers, so that each successive chamber encloses all previous chambers and thus the entire shell, is stated to occur in the morphospace where the ratio of the radius of a chamber to the distance from its midpoint and the midpoints of the preceding or succeeding chambers is greater than $(Kr - 1)^{-1}$. The geometric condition of a shell with completely involute chambers is essentially the opposite extreme from a "shell" with the chambers not touching one another at all.

Third, the region of uniserial shells, consisting of a series of chambers arranged in a straight line, is stated to occur in the morphospace where the value of $\phi$ approaches $\pi$, and that of $D$ is large (or the number of chambers per "whorl" is one, and the value of $Ky$ is large).

Fourth, the region of shells whose chambers are arranged in a logarithmic spiral in either two dimensions (planispiral condition, $Ky$ equal to zero) or three dimensions (helicospiral condition, $Ky$ greater than zero) is stated to occur in the morphospace between regions two and three discussed earlier.

Signes and colleagues (1993) do give computer-generated simulations of foraminiferal shells in their study, at least for "region four," which is the region containing the majority of planktic foraminiferal morphologies. Using their model, it would be possible to generate simulated foraminifera for the other three "regions" of their hypothetical theoretical morphospace as well and to construct a four-dimensional space by using the model parameters $Kt$, $\phi$, $Ky$, and $D$ as dimensional axes. They did not do this, however, and needless to say, they also did not complete the next step in a theoretical morphologic analysis, plotting of the positions of actual foraminiferal shells in the theoretical morphospace. Thus the foraminifera, a group with important paleoecologic and biostratigraphic significance in evolutionary studies and a group whose shells "exceed in architectural variety those of any other invertebrate group" (Brasier 1982a:1), still await a rigorous and complete theoretical morphologic analysis (see also Webb and Swan 1996).

Moving from unicellular forms of life to multicellular (and also to a group more closely related to us), the extinct graptolites also grew via a discrete process of successively adding thecae, producing a generally linear to curvilinear stipe. A rhabdosome (a colony of graptolite zooids) may consist of a single stipe but more often several stipes. Branching geometries of stipes in multiramous graptolite colonies were considered in chapter 3. Theoretical models of the stipe itself, in which the stipe is considered to be composed of "stacked cylinders" or "stacked cones," have been used to simulate hypothetical stipe morphologies and to construct a theoretical morphospace of graptolite stipe form.

Fortey (1983) modeled graptolite stipes as a series of stacked cylinders, where each theca is treated as a simple cylindrical tube. The cylinder is open at one end (the aperture) and closed at the other (where it encounters the common canal between the zooids and the stipe axis). A series of these cylinders are stacked side by side, all open in one direction and closed in the other, where the stipe axis is located. Five parameters are then used to define the geometry of the stipe:

1. $L$ = the "thecal length," the length of the cylinder, taken as the maximum distance from the stipe axis to the opening of the cylinder.
2. $t$ = the "thecal width," the diameter of the opening of the cylinder.
3. $\theta$ = the "thecal inclination," the angle between the axis of the cylinder and the stipe axis. If $\theta$ is equal to 90°, the cylinders project directly outward from the stipe axis, and all points around the margins of the openings of the cylinders are located at the same distance from the stipe axis. Thecae in real graptolite stipes are most often oriented at an angle to the stipe axis, which is less than 90°, giving the stipe the characteristic "saw-edge" or "zigzag" appearance in the side view. In real graptolite stipes, the ventral margin of the aperture is located at a greater distance from the stipe axis than the dorsal margin of the aperture. In the stacked cylinder model, all the cylinders are tipped over on their sides by the angle $\theta$,

resulting in some points of the margins of the openings of the cylinders being at a greater distance from the stipe axis than others (or to use the image of the proverbial row of dominos, the flat tops of the dominos all lie in a single plane when standing in a row but are stair-stepped when the row of dominos is tipped over).

4. $W$ = the "stipe width," the distance from the distalmost tip of the opening of the cylinder to the stipe axis.

5. $d$ = the "thecal spacing," the distance between the distalmost tips of the openings of successive cylinders.

If the thecae are a series of cylinders oriented at right angles to the stipe axis ($\theta$ equal to 90°), the width of the stipe ($W$) is then the same as the length of the thecae ($L$), and the spacing of the thecae ($d$) is equal to the width of the thecae ($t$). The algebraic relationship of $W$, $L$, and $\theta$ can be seen in figure 6.4, where $W$ and $L$ have been oriented as dimensional axes at right angles to each other, and the field of indices in the graph indicates the values of $\theta$ obtained by varying values of $W$ and $L$. Note that the index line for $\theta$ equal to 90° is the 45° diagonal, that is, the region in the morphospace where the value of $W$ is always equal to that of $L$.

The triangular region in the morphospace above the 45° diagonal is, in fact, another "geometrically impossible" region, similar to that seen in figure 4.5 for brachiopods and the "forbidden range" seen in figure 6.2 for foraminifera. Above the 45° diagonal in figure 6.4 would occur hypothetical graptolite stipes whose widths exceed the lengths of the thecal cylinders. In the model, the maximum stipe width occurs when $\theta$ is equal to 90° (when the axis of the thecal cylinder is oriented at right angles to the stipe axis), and 100% of the length of the theca contributes to stipe width. It is thus geometrically impossible in the model for stipe width to exceed thecal cylinder length. Biologically it is also impossible, as it would require the width of the common canal to exceed the length of the thecal tube. Interestingly, some graptolites do have $\theta$ values that exceed 90°, but this occurs only in forms in which the thecae are no longer straight cylinders (as in the model) but, rather, curved tubes that bend backward from their original growth orientation during ontogeny (Fortey 1983).

Note also in figure 6.4 that the values of the $\theta$ indices increase in a nonarithmetic manner as they approach the 45° diagonal. Thus for low values of $\theta$, small increases in $W$ produce large increases in $L$. Fortey (1983) argued that this geometric relationship is size limiting and that very few graptolites are found with stipe widths greater than 1.2 to 1.5

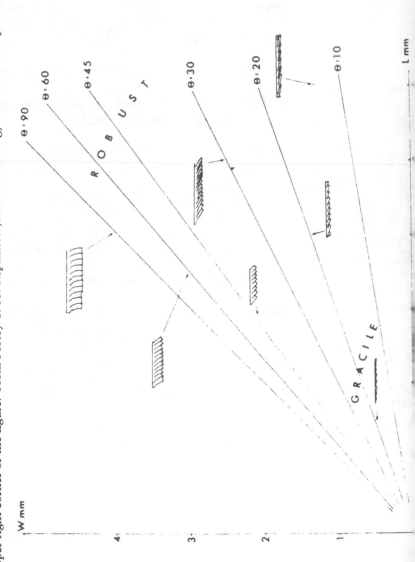

**FIGURE 6.4.** A section of Fortey's theoretical morphospace of graptolite stipe form (1983), illustrating hypothetical grapto- lites produced by varying the parameters thecal length (*L*, the x-axis) and stipe width (*W*, the y-axis). Distribution of values of the parameter thecal inclination (θ) are indicated by indices in the diagram. Illustrations of hypothetical stipe morphologies generated by the model are indicated for selected regions in the morphospace. Note that gracile stipe mor- phologies are found in the lower left corner of the morphospace, with stipe morphologies becoming more robust in form toward the upper right corner of the figure. *From Fortey 1983. Reprinted from Paleobiology and used with permission.*

mm if the thecal inclination is less than or equal to 20°. The geometric relationships of $W$, $L$, and $\theta$ allowed Fortey to specify regions of graptolite form in the theoretical morphospace. Graptolites with gracile stipe form occur in the lower left corner of the morphospace, having small stipe widths, thecal lengths, and low angles of thecal inclination. Graptolites with robust stipe form occur in the upper right corner of the morphospace, having large stipe widths, long thecae, and high angles of thecal inclination (figure 6.4). Similar plots of thecal width ($t$) versus thecal spacing ($d$) allowed Fortey to specify regions of thecal spacing relationships in the theoretical morphospace, from "loose" (stipes having low inclination angles with large thecal widths and spacings) to "dense" (stipes having high inclination angles with small thecal widths and spacings).

Even though the parameters $W$ and $L$ are shown as orthogonal axes in figure 6.4, Fortey (1983) points out in his paper that all the parameters of his theoretical morphospace are interrelated. Because the parameters are not algebraically independent of one another, they in fact do not form the orthogonal dimensions as are plotted in figure 6.4. Indeed, Fortey gives the algebraic nature of the relationship among the five parameters:

$$\sin \theta = W/L = t/d \qquad (6.3)$$

It is interesting that Fortey (1983) uses a triangular coordinate system when actually plotting measurement data from real graptolites in the theoretical morphospace of hypothetical graptolite stipe form (figure 6.5). Robust forms are found on the upper left side of the $L$-$W$–$D$ triangle, having approximately equal thecal lengths and stipe widths and small thecal spacings. Gracile forms are found on the upper right side of the triangle, having narrow stipes with long thecae and larger thecal spacings. Fortey terms this intermediate position "leptothecate." Graptolites in this region have stipes with long thecae but narrow widths and small thecal spacings.

Figure 6.5 gives the actual distribution of the stipes of 50 graptolite species in a $L$-$W$–$D$ section through the theoretical morphospace. The distribution forms a single cluster with stipe forms ranging from robust at one end-member to gracile at the other. Graptolites are not found, however, in regions where $L$ is greater than 80% or less than 40% of the total value of the three parameters. The modal graptolite, or peak of the distribution, occurs around the position of $W$ equal to 33%, $L$ equal to 50%, and $d$ equal to 17%. Thus the modal graptolite has a thecal spacing of about half the stipe width and a thecal length of about one and one-half of the stipe width.

Fortey (1983) does not discuss the functional significance, if any, of the observed distribution of the 50 graptolite species in the $L$-$W$-$D$ morphospace and so does not take the last step in a theoretical morphologic analysis, the functional analysis of the distribution of existent form in the theoretical morphospace of hypothetical form. He does discuss modifications of the model in which the cylinders representing thecae are replaced by cones and uses the model parameter $W$ to explore growth curves. With respect to the $L$-$W$-$D$ triangle, he concludes only that it "should prove a useful tool in plotting shifts in characters at the population level" (Fortey 1983:124).

Some of the geometric properties of the theoretical morphospace

FIGURE 6.5. The actual distribution of 50 graptolite species in the $L$-$W$-$d$ section of the theoretical morphospace of graptolite stipe form. Each corner represents 100%. Robust stipe morphologies occur near the $L$-$W$ axis, gracile morphologies near the $L$-$d$ axis, and leptothecate forms intermediately. *From Fortey 1983. Reprinted from* Paleobiology *and used with permission.*

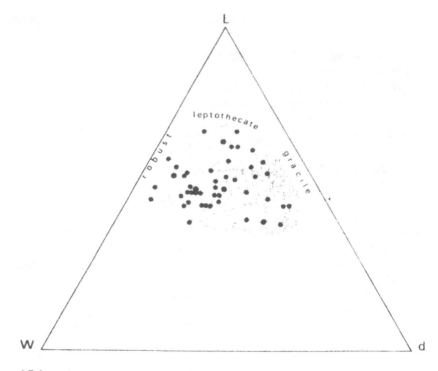

also remain unexplored. For example, not all the region inside the triangle given in figure 6.5 is actually available for potential graptolites. The extreme tips of the triangle are obviously impossible, as these correspond to a measurement condition of 100% for that particular parameter. It is not possible for a graptolite stipe to consist 100% of thecal length, with no thecal spacing or stipe width. Also, as stated earlier, stipe width may not exceed thecal cylinder length; thus the region above the 45° diagonal in figure 6.4 is a "geometrically impossible" region for graptolite stipes. In figure 6.5, this region corresponds to the region of the $L$-$W$-$D$ triangle, where the value of $W$ is greater than 50%. Fully a quarter of the $L$-$W$-$D$ triangle therefore is a geometrically impossible region (the lower left corner).

Before leaving this section of the chapter on theoretical morphospaces of chambers and cylinders, I must draw the reader's attention to the ingenious diagram of Niklas (1997b), in which a nine-dimensional hyperspace of plant cell morphology is represented in a three-dimensional format (figure 6.6)! By carefully studying the different permutations of cell geometries given in figure 6.6, in concert with imagining the range of morphologic variation implicit in each of the subdomain axes shown, the reader may visualize (or at least glimpse) what the total nine-dimensional space might actually look like.

## THEORETICAL MORPHOSPACES
## OF LAMINA AND LAYERS

The skeletons of massive reef corals are the product of growth and budding of the numerous and discrete polyps that make up the colony. Graus and Macintyre (1976) produced a computer simulation of coral colonial form by parameterizing the behavior of the individual corallites. Starting with a predetermined base, usually taken as a segment of a circle, the hypothetical corallites are positioned with specified distances between them and oriented vertically. These corallites then were allowed to grow vertically and bud horizontally (see also the model of discrete cell budding in the vertebrate limb morphogenesis of Ede and Law [1969]). The new corallites budded whenever intercorallite distance exceeded a specified value. Budding was frequent at the periphery of the colony but also occurred in the colonial skeleton as space was created by the curvature and divergence of the growing corallites across the differentially expanding surface of the colony. The growth surface of the colony was calculated in discrete steps, corresponding to the observed annular growth layers seen in actual colonial coral skeletons.

An additional twist was added to the simulation by making the growth rate of the individual corallites a function of the amount of light the corallite received (in nature the massive reef scleractinians are hermatypic; that is, their polyps contain symbiotic zooxanthellae that photosynthesize). Graus and Macintyre (1976) discovered that they could simulate many of the colonial coral forms seen in nature simply as a function of light intensity, which is usually a function of water depth. Corals located in very shallow water (1 m) tend to develop

FIGURE 6.6. Niklas's multicellular plant morphospace (1997b), in which aspects of a nine-dimensional hyperspace are ingeniously illustrated in a three-dimensional representation. The morphospace has three subdomains: cell form (lower right), tissue construction (upper right), and tissue tesselation (left). Because each of these three subdomains themselves have three dimensions (the three dimensions of the cell form subdomain are shown in the lower right), the total morphospace is nine dimensional. Such a hyperspace cannot be faithfully rendered in three dimensions, but the choice of geometric combinations given in the figure allows the viewer to visualize what such a morphospace might actually look like. *Artwork courtesy of K. J. Niklas. Copyright © 1997 by the University of Chicago and used with permission.*

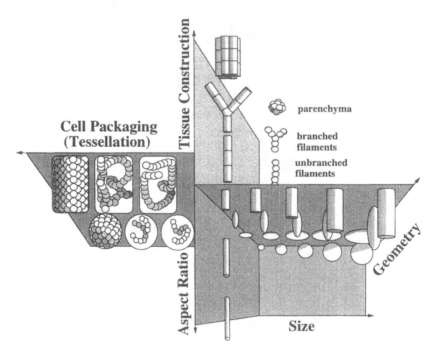

hemispherical forms, whereas those in deeper water (20 m) produce tall, pillarlike colonies. Corals in still deeper (and darker) habitats tend to spread out horizontally rather than growing vertically, producing saucer-shaped colonies. Each of these characteristic forms could be produced by computer simulation simply by varying the growth rates of the individual corallites as a function of light intensity. Thus the observed variation in coral morphology seen in nature could simply be a function of the amount of light received by the colony and not be due to internally programmed genetic differences among different colony forms.

Although Graus and Macintyre (1976) did compare their computer-simulated forms with actual coral colonies, they did not take the further step of creating a theoretical morphospace of colonial coral form, although this could be done using the parameters of the model they used. Interestingly, the addition of ecologic dimensions, namely, light intensity and water depth, to such a space would actually lead to the creation of a "theoretical design space" rather than a theoretical morphospace (see the discussion of theoretical design spaces in chapter 2).

A theoretical morphospace of massive laminate form was created by Kershaw and Riding (1978), but for stromatoporoids rather than corals. The skeleton of stromatoporoids is a reticulum composed of discrete "latilaminae," banded structures consisting of chambers surrounded by vertical pillars of skeletal material connected by thin horizontal laminae, radial processes, or tabulae between chambers. In general, soft tissue occurs only in the upper layers of the skeleton, although the total skeleton may be composed of numerous latilaminae. Rather than attempting to simulate the complex internal structure of the latilaminae that comprise the skeleton, Kershaw and Riding (1978) focused on the gross external morphology of the stromatoporoid, specifically the shape of the skeleton in cross section. Their model of stromatoporoid morphology contains four parameters:

1. $B$ = the "basal length" of the cross section through the stromatoporoid, taken simply as the distance from one side of the coenosteum to the other across its base.
2. $V$ = the "vertical height" of the cross section, taken from the midpoint of the basal length and at a right angle to the basal dimension.
3. $D$ = the "diagonal distance" of the cross section, a line taken from the midpoint of the basal length and at an acute angle to the vertical height, extending from the base of the cross section to its outer periphery. In the model, the shape of the

cross section is held symmetrical about its vertical height; thus $D$ taken to the left of $V$ and $D$ taken to the right of $V$ are the same length.

4. $\theta$ = the angle between $V$ and $D$.

If the reader visualizes a cross section through a stromatoporoid skeleton as a simple hemicircle, then the diameter of the hemicircle is $B$; the radius at right angles to the diameter is $V$; and $D$ is any other two radii located at the acute angle $\theta$ on either side of $V$.

A three-dimensional section through the four-dimensional theoretical morphospace of stromatoporoid cross sectional form is given in figure 6.7, where the parameter $\theta$ is held constant at 25° and the parameters $V$, $B$, and $D$ are allowed to vary. Kershaw and Riding (1978) chose to arrange their parameters in a triangular coordinate format, similar to that used by Fortey (1983) in figure 6.5. The corners of the triangle represent the 100% condition of that particular parameter, and the effect of covariation in any two parameters is shown in the graphical simulations arranged along the edges of the triangle. At the lower left corner, for example, is illustrated the condition in which $B$ is 100%, which is a simple horizontal line. Progressing up the left side of the triangle shows the effect of increasing the magnitude of $V$ relative to $B$, until at the apex $V$ alone is present in a simple vertical line. Likewise at the lower right corner, $D$ alone is present and moving up the right side of the triangle illustrates the progressive overtaking of the diagonal dimension by the vertical.

In the spirit of Raup (1966), who sketched out a predicted distribution of selected molluscan groups in the $W$–$D$–$T$ theoretical morphospace of univalved shell form on the basis of the simulated morphologies alone (figure 4.3), Kershaw and Riding (1978) drew in the $V$–$B$–$D$ theoretical morphospace their predicted distribution of three common types of stromatoporoid skeletons: laminar, domical, and bulbous (figure 6.7). They then went further, however, and plotted the distribution of 60 specimens of actual Silurian stromatoporoids, from two separate field locations in Gotland, in the theoretical morphospace (figure 6.8). The distribution of actual stromatoporoids (figure 6.8) lies in the predicted distribution of common skeletal types (figure 6.7). Interesting locality differences can also be seen in the morphospace: the range of morphologic variability, or biodisparity, is less in one locality (solid circles enclosed in the dashed line in figure 6.8) than in the other (open circles in the figure). In the latter site, stromatoporoid morphologies are found that are both more the laminar type (toward the lower left corner of the morphospace) and more the bulbous type (toward the upper right side of the triangle) than the former.

It is important that entirely conceivable stromatoporoid skeletons exist in the theoretical morphospace of hypothetical forms, but these nevertheless do not actually appear in nature. In neither the distribution of measurements from actual stromatoporoids (figure 6.8) nor the predicted field of distribution of common skeletal types (figure 6.7) is

FIGURE 6.7. Kershaw and Riding's *V–B–D* theoretical morphospace of stromatoporoid form (1978). Illustrated are the outlines of hypothetical stromatoporoid cross sections produced by variation in the parameters vertical dimension, basal dimension, and diagonal dimension. The estimated distribution of common stromatoporoid forms in the theoretical morphospace is indicated by the dashed outlines, which enclose the expected fields of laminar (L), domical (D), and bulbous (B) stromatoporoid morphotypes. *Reprinted from* Parameterization of stromatoporoid shape *by S. Kershaw and R. Riding from Lethaia 1978, volume 11, pages 233–242, by permission of Scandinavian University Press. Artwork courtesy of S. Kershaw.*

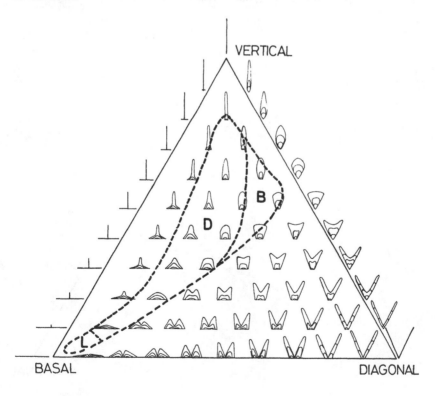

the lower right corner of the theoretical morphospace seen to be occupied. This region is theoretically possible but actually is empty. Kershaw and Riding (1978) note that in general, stromatoporoids appear not to have utilized forms with concavities in the upper surface. They argue that the usage of exclusively convex upper surfaces in the colonial skeletons of these filter-feeding organisms is due to sediment avoidance, an adaptation to shed sediment or at least to hinder its accumulation on the colony.

Kershaw and Riding's study (1978) completes all the steps of a theoretical morphospace analysis: from constructing a theoretical morpho-

FIGURE 6.8. The actual distribution of 60 specimens of Silurian stromatoporoids in Kershaw and Riding's (1978) theoretical morphospace of stromatoporoid cross-sectional form. The stromatoporoid specimens were taken from two separate field localities in Gotland: solid circles indicate the Upper Visby marlstone, and open circles the Hemse biostrome. *Reprinted from* Parameterization of stromatoporoid shape *by S. Kershaw and R. Riding from Lethaia 1978, volume 11, pages 233–242, by permission of Scandinavian University Press. Artwork courtesy of S. Kershaw.*

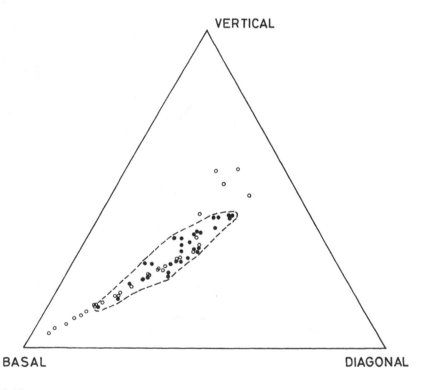

space to the exploring the distribution of actual form in that morpho-space and subsequently analyzing the functional significance of both existent and nonexistent morphology. In sum, it is an excellent study, so I find it disturbing that they apologize in their study for the construction of the theoretical morphospace itself:

> We found that the array of shapes resulting from our scheme is sufficiently simple to be drawn reasonably accurately by hand. Consequently, we have neither produced nor displayed our scheme in a sophisticated manner and are aware of the contrast between our methods and those of some other parameterizations, such as Raup's classic work on coiled shells. . . . We hope that the simplicity of our methods will be justified by the ease with which the resulting arrays can be used to elucidate features of stromatoporoid morphology. (Kershaw and Riding 1978:237)

This apology again reflects the unfortunate misconception concerning theoretical morphology we examined earlier in chapter 2, namely, that all theoretical morphospaces must be the products of complex mathematics and sophisticated computer graphics. I hope this book corrects that misconception.

Kershaw and Riding (1978) concluded that stromatoporoids have not utilized skeletal forms with concavities in the upper surface (although these are possible, see figure 6.7) and that this is probably due to sediment avoidance. This idea was pursued further by Swan and Kershaw (1994), but by using a "theoretical design space" rather than a theoretical morphospace. They designed a model using discrete "pixels," arranged in a reticular grid, to simulate the lamination found in stromatoporoid skeletons. Starting with a seed pixel, adjacent cells in the grid surrounding the seed pixel are either "switched on" or left "switched off" according to a probability function, producing roughly concentric growth in a horizontal plane. Growth in the vertical dimension is accomplished in a similar fashion, in which the vertical probability function simulates negative geotropism, that is, the tendency to grow upward for purposes of avoidance. Negative geotropism is commonly found in marine benthic organisms that must grow vertically to avoid being buried by sediment, to escape competition from adjacent neighboring organisms, or both. To further simulate the stimulus for negative geotropism, Swan and Kershaw (1994) added the ecologic dimension of sedimentation to the model. Sediment input was varied in terms of three possible parameters: (1) the amount of sediment in each sedimentation event, (2) the interval of time between

**163**

each sedimentation event, and (3) the length of the hiatus of time before sedimentation during which the initial establishment of the stromatoporoid occurred.

An example of some of Swan and Kershaw's (1994) simulations is
given in figure 6.9. Note that although the simulations in the figure
resemble stromatoporoid morphologies, the dimensions of the space
are not morphological but ecological. The y-axis is a measure of sedimentation rate, and the x-axis is a measure of the frequency of sedimentation. Along the bottom margin of the figure, where the value of
the y-axis is zero, are simulations of stromatoporoids produced simply
according to the values of the probability functions determining their
horizontal and vertical growth. Alterations induced in these simple
forms by adding sediment influx can be seen in the remainder of the
stromatoporoid simulations. The top left simulation shows a stromatoporoid that must contend with a large amount of sediment that occurs in numerous small increments; the top right simulation shows a
stromatoporoid that must deal with the same amount of sediment but
that arrives in larger increments less frequently. In the region of the
space above the top row of simulations, the sedimentation rate exceeds the growth rate of the stromatoporoid simulations, so that hypothetical stromatoporoids in this region are smothered soon after they
begin to grow. Comparison of real stromatoporoid morphologies with
the hypothetical ones seen in figure 6.9 led Swan and Kershaw to conclude that the response to differing regimes of sedimentation is indeed
a major factor in determining stromatoporoid shape.

The morphospace shown in figure 6.9 contains both morphologic
parameters and ecologic parameters. It thus meets the defining criteria
of a "theoretical design space" (see the discussion in chapter 2) and
could be described as a "theoretical design space of negative geotropism in stromatoporoid form." And as noted at the beginning of this
section, a similar "theoretical design space of phototropism in coral
form" could likewise be created using Graus and Macintyre's (1976)
model.

The laminae or layers found in stromatolites are not the product of
discrete polyp budding patterns in a colony as for corals, nor are they
composed of numerous discrete vertical pillars and horizontal tabulae
or laminae as for stromatoporoids. They are actually more accretionary in nature, being produced by alternating layers of cyanobacterial
mats and sediment drapes. The morphologies of many stromatolites,
however, resemble those modeled for stromatoporoids in figure 6.9. It
is also equally likely that the response to differing regimes of sedimentation is a major factor in determining stromatolite shape, as well

as the amount of light reaching the photosynthetic cyanobacteria at varying water depths. Thus the models developed for corals by Graus and Macintyre (1976), and for stromatoporoids by Kershaw and Riding (1978) and Swan and Kershaw (1994), may be easily applied to stromatolitic morphologies as well.

A simple theoretical morphospace of stromatolite form was produced by Hofmann (1994), although termed a "laminosity plot" in the original study. Hofmann used the vertical height, horizontal width, and arcuate length of a lamina, as seen in cross section, to create the

**FIGURE 6.9.** A "theoretical design space of negative geotropism and stromatoporoid form," created by Swan and Kershaw (1994). Note that although stromatoporoid morphologies are simulated in the figure, the dimensions of the space are not morphological but ecological. The *y*-axis is a measure of sedimentation rate, and the *x*-axis is a measure of the frequency of sedimentation; see the text for discussion. *From Swan and Kershaw 1994. Reprinted with permission of the Palaeontological Association, London.*

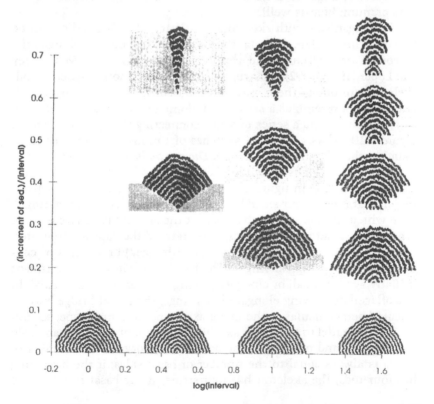

three dimensions of horizontal laminosity, vertical laminosity, and profile flatness in the laminosity plot. Therefore the laminosity plot is similar to the $V–B–D$ theoretical morphospace of stromatoporoid form (figure 6.7), but without the diagonal dimension. The laminosity plot can be used to create both existent and nonexistent stromatolite laminar forms, and so I include it here as a theoretical morphospace.

## THEORETICAL MORPHOSPACES OF RODS AND PLATES

Many organisms possess skeletons composed of many interlinked discrete elements, such as rods and plates. The human skeleton itself can be considered to consist of a series of rods that are more or less straight (the humeri, radii, ulnae, femora, tibiae, and fibulae of our limbs) or curved (the ribs) or compressed (the vertebrae. An aside: For a promising first step in creating a theoretical morphospace of vertebrate skeletal form, see Van Valkenburgh [1985, 1988]). Organisms with skeletons consisting of interlocking plates range from the microscopic dinoflagellates to large sea urchins (and we could perhaps include the human cranium here as well).

Microorganisms with skeletons consisting of rods of silica can be found in three different phyla: the Chrysophyta (silicoflagellates), the Sarcodina (radiolarians), and the Pyrrophyta (ebridians). McCartney and Loper (1989, 1992) constructed a theoretical morphospace of rod-skeletons for one of these groups, the silicoflagellates. A silicoflagellate skeleton resembles a small open dome or pyramid, consisting of an apex, a base, and a series of struts connecting the apex to the base. Figure 6.10 shows a series of sketches of various morphotypes to be found in the silicoflagellate genera *Dictyocha* (sketches A through D in figure 6.10) and *Distephanus* (sketches E and F). The medusid morphotype (sketch D in figure 6.10) is very much like a tiny open pyramid in that it has a four-sided base of connected rods (the "basal ring"), from which rise four struts that meet at the apex. Often there is some type of rod structure at the apex, aptly termed the "apical structure," which is either a single rod (the "apical bridge") or a series of connected rods (the "apical ring"). The basal ring can be either equant (radially symmetrical) or elongate (having a major and minor axis). In silicoflagellates having elongate basal rings, the apical bridge may be oriented perpendicular to the major axis (asperid morphotype, see figure 6.10), parallel to the major axis (fibulid morphotype), or at an angle to the major and minor axes (aculeatid morphotype). The number of struts is always equal to the number of basal sides: if the basal ring has four rods, the skeleton has four struts; if the basal ring has five rods, the skeleton has five struts; and so on. If an apical ring is present, it has the same number of rods as the basal ring (see sketches E and F in figure 6.10).

McCartney and Loper's (1989) theoretical morphospace of silico-flagellate rod-skeletons has four parameters:

1. $\alpha$ = the "aspect ratio" of the basal ring, measured as the length of the base axis aligned with the bridge axis divided by the length of the other base axis.
2. $\theta$ = the "bridge rotation" angle, measured with respect to the basal axis with which the bridge axis is most closely aligned.
3. $\beta$ = the "relative length" of the bridge, the length of the bridge divided by the length of the basal axis with which it is aligned.
4. $\phi$ = the "relative height" of the bridge, the height of the bridge attachments divided by the height of the apex of the skeleton.

Figure 6.11 shows a two-dimensional slice through McCartney and Loper's (1989) four-dimensional theoretical morphospace. Note that fibulid morphotypes are produced when values of the aspect ratio ($\alpha$) are greater than 1.0 and that asperid morphotypes produced when $\alpha$ is less than 1.0 (simulations arranged along the right margin of figure 6.11). If $\alpha$ is equal to 1.0, the basal ring is equant and has no major or minor axes. The effect of varying the value of the bridge rotation angle ($\theta$) is shown in the simulations arranged along the top margin of figure 6.11 for both fibulid (top row) and asperid (bottom row) morphotypes. The top row of simulations occurs in the region of the morphospace where $\alpha$ is greater than 1.0, and the bottom row of simulations occurs in the morphospace where $\alpha$ is less than 1.0.

Instead of the silicoflagellate simulations (which are arranged around the margins), a series of contour lines are shown in the theoretical morphospace itself (figure 6.11). These contours give values of the "relative apical surface area" found in various silicoflagellate skeletons in the theoretical morphospace. The relative apical surface area of a silicoflagellate skeleton is defined as the ratio of apical surface area divided by the basal ring area (McCartney and Loper 1989:287). The lowest values of relative apical surface area are found in the lower left margin of the morphospace, where $\theta$ is zero and $\alpha$ is around 0.85 (figure 6.11). Thus the *Dictyocha* morphotype having the minimum apical surface area, in the boundaries of the theoretical morphospace, has no bridge rotation and is asperid in form. Rotation of the bridge

167

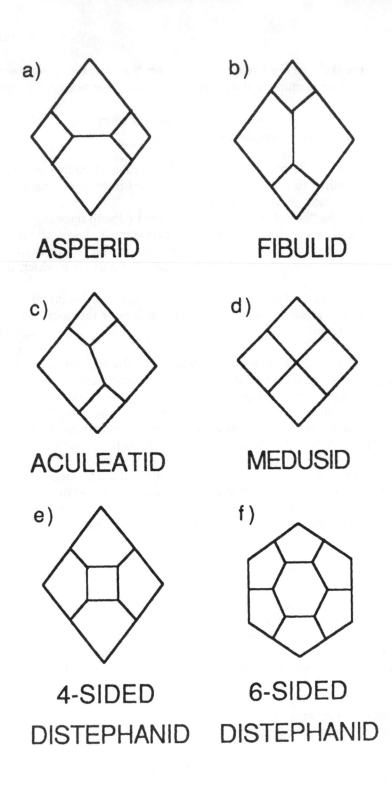

a) ASPERID

b) FIBULID

c) ACULEATID

d) MEDUSID

e) 4-SIDED DISTEPHANID

f) 6-SIDED DISTEPHANID

(producing aculeatid forms) to larger and larger angles progressively produces larger apical surface areas in the skeleton (figure 6.11). Similar simulations with the *Distephanus* morphotypes reveals that a minimum apical surface area is produced in these forms when θ is zero (no bridge rotation) and α is one (i.e., the basal ring is equant and has no major and minor axes).

McCartney and Loper (1989) and McCartney, Ernisse, and Loper (1994) exhaustively explored the geometric relationship of two other aspects of silicoflagellate structure in the range of hypothetical skeletal forms found in their theoretical morphospace. These two other measures are the relative volume of the skeleton and the relative length of the skeletal elements. They found that simulated morphotypes that minimize the relative apical surface area versus a given relative volume consistently produced forms that are frequently found in nature, whereas simulated morphotypes that minimize the relative length of the skeleton elements versus the relative volume produced forms that are less commonly found in nature or forms that do not naturally occur at all. This last finding is interesting, as silica is currently undersaturated in the oceans and is thus a limiting resource. A skeleton that possesses minimum lengths of the structural elements needs the minimum amount of silica to construct those elements. Skeletons with minimum rod lengths are also lighter so presumably have less problem maintaining buoyancy in the water column. The conclusion of McCartney and Loper's (1989) theoretical morphologic analysis of silicoflagellate skeletal form is, however, that silicoflagellates are primarily minimizing apical surface area and that silica conservation is only a secondary factor in skeletal construction. They consider that the skeleton's primary function is to support the apical cell boundary in a shape that minimizes its surface area and that only when this primary function is achieved does the silicoflagellate utilize geometries that use as little silica as possible in achieving that goal.

The functional conclusions of the McCartney and Loper (1989) and

FIGURE 6.10. "Stick-figure" sketches of common silicoflagellate skeletal morphologies in apical view. Orientation: the center of the sketches is "up" and contains the skeleton's apical bridge (sketches A through C), the apical ring (sketches E and F), or the simple fusion of the struts (sketch D). The outer margin of the sketches is "down" and represents the basal ring of the skeleton. The lines connecting the center and margin of the sketches are the struts of the skeleton, extending from the apex to the base. Morphologic terms used for the different skeletal types are given in the figure. *From McCartney and Loper 1989. Reprinted from* Paleobiology *and used with permission.*

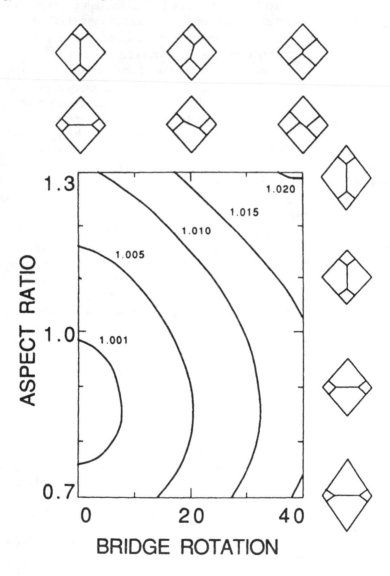

McCartney, Ernisse, and Loper (1994) studies are based on comparisons of real silicoflagellate skeletons with hypothetical ones produced by varying the model parameters. Although measurements taken from 1,000 silicoflagellate specimens are mentioned in the text of the studies, McCartney and Loper (1989) and McCartney, Ernisse, and Loper (1994) do not give plots of those measurements in the theoretical morphospace format given in figure 6.11. McCartney and Loper (1989) in effect argued that the contour surface of relative apical surface area seen in figure 6.11 is an adaptive hill, and so it would be extremely interesting to see the density distribution of actual measurement points taken from existent silicoflagellates in the α–θ space shown in figure 6.11. Presumably the great majority of points would fall in the 1.001 contour in figure 6.11, with fewer and fewer points scattered in the theoretical morphospace as the contour magnitudes of relative apical surface area increase.

All in all, however, McCartney and Loper (1989, 1992) created a theoretical morphospace of silicoflagellate form. A theoretical morphospace of the more complex rod-skeletons of the microscopic Radiolaria has yet to be created. Other macroscopic organisms that produce skeletons of interlinked rods, most notably the Porifera, also await a theoretical morphologic analysis.

Teeth are discrete skeletal elements that might also be thought of as rods, or at least as conical rods. Theoretical morphogenetic models of the generation of teeth in vertebrates and the teethlike denticle scales of sharks were produced by Weishampel (1991) and Reif (1980) and will be considered in more detail in chapter 10. Reif (1980) went further, however, and used his model to create a semiquantitative theoretical morphospace that could be termed an "elasmobranch dermal skeleton space" or a "sharkskin space." In this theoretical morphospace, Reif (1980) outlined 18 regions, or "squamation types," one of which is biologically impossible (Reif calls it a "nonsense type"). Thus the dimensional combinations of Reif's theoretical morphospace can also create regions similar to the "geometrically impossible" zone for brachiopods (figure 5.5) and the "forbidden range" for foraminifera (figure 6.2), which we examined earlier. Even though the dimensions of Reif's theoretical morphospace are semiquantitative and no actual measurements are given., Reif did outline the expected distribution of major existent shark types in the morphospace and also a region that is expected to be unoccupied in nature (a region characterized by both high growth rates and high shedding rates of the scales). Reif's sharkskin morphospace is an excellent beginning and holds promise for future development and quantification.

The theoretical morphologic analysis of organisms with skeletons consisting of discrete plates, like that of skeletons of interlinked rods, is also in its infancy. This fact is curious in that the founder of theoretical morphology himself, D. M. Raup, designed a theoretical model of plate morphogenesis in echinoids as early as 1968. The main skeleton of an echinoid (excluding spines and so on) consists of numerous columns of plates (the ambulacral and interambulacral columns) that radiate out from the apical system (the dorsal region of the skeleton). Raup (1968) argued that echinoid skeleton shape is fundamentally a function of plate development and configuration. In that early study, Raup assumed that the geometric pattern of the plate mosaic seen in echinoid ambulacral and interambulacral columns was simply the result of close-packing, a geometric constraint underlying such disparate phenomena as the hexagonal shape of cells seen in honeycombs (Thompson 1917, 1942) to the ubiquity of the angle 120° in organic geometries (Stevens 1974). In the specific case of echinoids, the close-packing assumption is that the perimeter of the plate is minimized with respect to the surface area of the plate. If the boundary shapes of the plates are determined by close-packing constraints, what determines the plates' position and size?

The plates in an echinoid skeleton are produced at the apical system and then progressively migrate in a column away from the dorsal (adapical) region to the ventral (adoral) region of the skeleton as new plates are produced at the apical system. The rate of migration of a given plate away from the adapical region is thus a function of the rate of supply of new plates at the apical system. In his morphogenetic model, Raup (1968) set the rate of plate production as a logistic function of time. That is, the rate of plate production at the apical system is initially low, increases to a maximum, and thereafter decreases in time. Variation in growth patterns between different echinoid individuals was simulated by utilizing differing parts of the logistic function in the simulations.

The position of a plate in the plate column is a function of its rate of migration, which in the model is the logistic function of the rate of plate production. The size of the plate in the model is taken as the magnitude of its meridional (dorso-ventral) axis. Raup (1968) did not attempt the added complexity of modeling growth in the lateral axis of the plates but, rather, simply fit the width of the ambulacral column of plates to an arbitrary segment of a circle to simulate the curvature seen in an echinoid skeleton. The growth increment in the meridional axis of a plate was set as a parabolic function of its distance from the apical system, where the two positive limbs of the parabola occur at the adapical and adoral regions and the vertex of the parabola is near the ambitus in between. That is, growth increments in plate size are initially high (at the apical system), decrease in magnitude as the ambitus is approached, and then once again increase in magnitude (approaching the adoral region). Negative growth was also allowed for, as plates in echinoid skeletons are often resorbed (and hence decrease in size) during growth, rather than simply becoming ever larger. In the model, negative growth is simulated by moving the vertex of the parabolic growth function to the region of negative numbers along the growth increment axis.

Two entirely hypothetical, computer-produced, "echinoid ambulacral columns" of plates are given in figure 6.12, generated by Raup's (1968) theoretical morphogenetic model of plate growth. The two theoretical ambulacral columns were produced by modifying the rate of plate production and the magnitude of plate growth increment in the model. In the simulation on the left (figure 6.12A), the rate of plate production was held constant by using only the central part of the logistic growth curve, and in the simulation on the right (figure 6.12B), the rate of plate production was allowed to increase ontogenetically by using the initial part of the logistic growth curve. In both simulations, the magnitude of the plate growth increment followed a parabolic function of decreasing, negative (resorbtion), then increasing increment sizes from top to bottom in the column of plates. The morphogenetic result was an ambulacral column where either maximum plate sizes (figure 6.12A) or minimum plate sizes (figure 6.12B) occur at the ambitus.

Raup (1968) was interested in producing a theoretical morphogenetic model of echinoid plate growth, not in producing a theoretical morphospace of hypothetical echinoid form. It would be possible, however, to construct a "theoretical morphospace of plate columns" by using the morphogenetic model he created. The simulations given in figure 6.12 could in essence be used to define the two end-members of a dimensional axis of "plate size at the ambitus," from maximum to minimum, with other permutations in between produced by further experimentation with the morphogenetic model. Creating a theoretical morphospace of echinoid form "from the plate on up" would require tackling also the determinants of lateral plate magnitude and the generation of multiple columns of plates.

In contrast to the formal mathematical modeling of ambulacral plate growth seen in Raup's (1968) study, Strathman (1975) conducted a much simpler but highly interesting "thought-experiment" in theoretical ambulacral form in echinoderms. He began by simply considering two questions: (1) What is the function of the ambulacra? and (2)

**FIGURE 6.12.** Theoretical echinoid ambulacral columns produced by modifying the rate of plate production and the plate growth increment in the morphogenetic model. In the simulation on the left (figure 6.12A), the rate of plate production is held constant, and in the simulation on the right (figure 6.12B), the rate of plate production increases with time. In both simulations, the magnitude of plate growth increment follows a parabolic function of decreasing, negative (resorbtion), then increasing increment sizes from top to bottom in the column of plates. The result is an ambulacral column in which either maximum plate sizes occur at the ambitus (figure 6.12A) or minimum plate sizes occur at the ambitus (figure 6.12B). *From Raup 1968. Used with permission of SEPM—Society for Sedimentary Geology (formerly Society of Economic Paleontology and Mineralogy).*

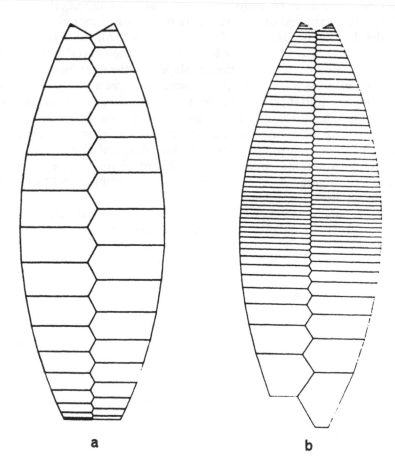

a                                        b

How do the ambulacra grow? He reasoned that because the food-gathering ability of an ambulacrum is a function of its surface area and the food requirements of the echinoderm is a function of its body volume, it would be advantageous to the echinoderm to somehow differentially increase the surface area of the ambulacra with growth. Using simple hand-drawn graphics, Strathman explored the various ways in which ambulacral surface area might be increased given the growth constraints of the echinoderm water vascular system. In the process, he created hypothetical ambulacral systems and compared those systems with ones actually occurring in echinoderms (either graphically or verbally). He even created a nonexistent echinoderm, one in which food-gathering ability is maximized yet nonetheless with a morphology not found in nature! All the elements of theoretical morphospace construction are present in the paper, although Strathman did not formally combine his various reasoning "parameters" into a hyperspace format.

An opposite approach to the theoretical morphologic analysis of echinoderm form was taken by Dafni (1986) and Ellers (1993). These authors focused solely on the gross morphology of the echinoderm skeleton rather than its constituent plates, much as Kershaw and Riding (1978) focused on the gross morphology of stromatoporoids rather than the intricacies of the many pillars and laminae making up the stromatoporoid skeleton. The studies by both Dafni (1986) and Ellers (1993) concern the theoretical morphology of regular echinoids.

Dafni (1986) was inspired by the analogy drawn by Thompson (1917) on the morphologic similarity of the shape of a regular echinoid skeleton and the shape of a drop of liquid resting on a smooth planar surface. He thereby conducted a series of experiments to test the idea that biomechanical constraints, rather than phylogenetic inheritance or direct genetic control, is the prime determinant of regular echinoid morphology. His research generally supports the biomechanical model of an echinoid skeleton as a "tensional sphere, the bubble-like pneu" (Dafni 1986:156). And even though the plates of the echinoid skeleton are rigid on a momentary basis, they behave elastically on the time scale of the animal's lifetime growth.

The Thompsonian model of a sea urchin skeleton as a "tensional sphere" was further explored by Ellers (1993), who argued that "an urchin's shape, despite complex details of plate growth, is thus determined by a force balance at each point in the skeleton" and that "despite their apparent rigidity, over developmental time spans urchin skeletons are inferred to deform in a manner similar to a liquid droplet" (Ellers 1993:123, 126). He then developed a "membrane model"

175

of form in regular echinoids, in which the skeleton is directly analogous to a liquid droplet. The shape of the droplet is a function of the balance between the surface tension and the internal pressure of the droplet and can be described in terms of three parameters:

1. $r$ = the "apical curvature" of the droplet, which is measured as the radius of curvature of the droplet at its apex.
2. $p_a$ = the "apical pressure" of the droplet, the internal pressure directed upward from the apex of the droplet.
3. $\gamma$ = the "gradient of pressure with depth" in the droplet, from its apex to its base.

To create a theoretical morphospace of echinoid form (figure 6.13), the three parameters were reduced to two by using the ratio value $\gamma/p_a$ rather than the separate values. A series of droplet shapes were created by systematically varying the parameter values $r$ and $\gamma/p_a$, and these shapes were compared with the meridional profiles of actual echinoids. The heights and diameters of the echinoids that matched the simulations were measured and mapped as two fields in the two-dimensional morphospace defined by the parameter axes of $r$ and $\gamma/p_a$. One field of values shows the range in skeletal heights produced by differing values of $r$ and $\gamma/p_a$, and another shows the range of values of the ratio of height to diameter (figure 6.13).

Ellers (1993) noted that echinoids often do not grow isometrically, in that some become taller with growth whereas others become flatter. He measured the skeletal heights and diameters of an ontogenetic series of individuals belonging to two species, *Stronglyocentrotus purpuratus* and *S. franciscanus*, and plotted these measurements in the theoretical morphospace (figure 6.13). He was able thus to illustrate the actual growth trajectories of form in the theoretical morphospace, in which one species (*S. purpuratus*) exhibited a marked change in the ratio of height to diameter during its ontogeny and the other (*S. franciscanus*) showed more marked change in height alone (figure 6.13).

If only height and diameter measurements are needed to plot individuals in the theoretical morphospace, why not simply use these measurements to define the axes of a morphospace, rather than the more exotic parameters of $r$ and $\gamma/p_a$? One answer to that question is that one cannot simulate the shapes seen in echinoid skeletons by using just the linear parameters height and diameter, whereas one can create droplet-shaped outlines that match those seen in a meridional profile of an echinoid skeleton with the tensional-membrane model. The visual presentation of the theoretical morphospace given in figure

6.13 would have been greatly enhanced if Ellers (1993) had given simulated meridional profiles of hypothetical echinoids as part of the figure, indicating for each simulation where it would fall in the field of
the morphospace (i.e., indicating which permutation of $r$ and $\gamma/p_a$ values correspond to each simulation). He does provide such simulated

FIGURE 6.13. Ellers's $r$-$\gamma/p_a$ theoretical morphospace of regular echinoid
form (1993). The dimensional axes are apical curvature of the skeleton ($r$)
and the ratio between the dorso-ventral pressure gradient in the skeleton
($\gamma$) and the apical pressure of the skeleton ($p_a$). Variation in these parameters is used to simulate hypothetical echinoid meridional profiles; see the
text for discussion. Indices in the morphospace give typical values of
height and the ratio of height to diameter seen in echinoid skeletons that
match the corresponding $r$ and $\gamma/p_a$ values. Measurements taken from an
ontogenetic series of individuals of two species, *Stronglyocentrotus purpuratus* (solid dots) and *S. franciscanus* (open circles), show the allometric
growth trajectories of these two species in the morphospace (from the upper left to the lower right for both species). *Artwork courtesy of O. Ellers.*

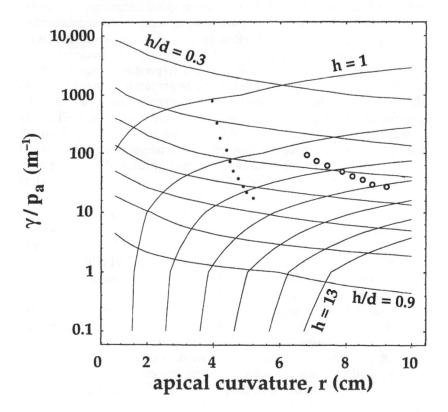

profiles elsewhere in the paper, so this could easily have been done (although given the small printed size of figure 6.13 in the original publication, I suspect editorially imposed space constraints may have prevented a more elaborate figure).

Ellers (1993) also considered figure 6.13 to be a "developmental" morphospace rather than a "theoretical" morphospace. He defined a "developmental" morphospace as one in which a developmental model (i.e., a morphogenetic model) is used to "determine the parameters that define a morphospace," a concept that he considered to be new and different from a theoretical morphospace, as he explicitly stated (1993:127–128): "Developmental morphospace is a new tool for comparative biology and for studying evolutionary dynamics. . . . It borrows the concept of morphospace [Raup 1966] but adds meaning to the axes that define the morphospace."

I consider there to be no fundamental difference between the $r$–$\gamma/p_a$ morphospace in figure 6.13 and Raup's $W$–$D$–$T$ morphospace in figure 4.3. If one considers that a "*theoretical* morphospace" is a hyperspace constructed using *theoretical morphologic parameters* to define the dimensions of the space and that an "*empirical* morphospace" is a hyperspace constructed using *empirical morphologic measurements* to define the dimensions of the space, then one must conclude that a "*developmental* morphospace" is a hyperspace constructed using *developmental morphologic parameters and/or measurements themselves* to define the dimensions of the space. And this is not the case for the *parameters* of the $r$–$\gamma/p_a$ morphospace in figure 6.13. The parameters of Ellers's (1993) model and Raup's (1966) model both are the products of morphogenetic models, a tensional-membrane model of morphogenesis for echinoids and a logarithmic spiral model of morphogenesis for gastropods and other univalved molluscs. The "radius of apical curvature," $r$, in figure 6.13 is just as much a geometric parameter as the "distance from the coiling axis," $D$, in figure 4.3. And parameters representing a "pressure gradient," $\gamma$, or an "expansion rate," $W$, are both firmly rooted in physics and equally as "meaningful" to an echinoid or a gastropod. Because the echinoid's coelomic pressure is an aspect of its morphology, $\gamma/p_a$ is a morphologic parameter and not an external environmental factor (if it were, the morphospace would be a "design space," as in figure 6.9). Hypothetical echinoid morphologies, potentially existent or nonexistent in nature, can be produced by varying the parameters of the $r$–$\gamma/p_a$ morphospace without any reference to measurement data taken from actual morphologies; thus the morphospace meets the criteria of a theoretical morphospace (as discussed in chapter 2), and I include it as such here.

Moore and Ellers (1993) also created what they term a "functional morphospace" for aspects of sand dollar form, a morphospace that converges on an adaptive landscape, as discussed in chapter 2.

For echinoderms, a very diverse group of organisms with skeletons consisting of discrete plates, we therefore have only one existent theoretical morphospace (for regular echinoids only) and a potential "theoretical morphospace of plate columns" that could be constructed using Raup's (1968) plate morphogenetic model. Theoretical morphospaces of crinoid morphology are being developed, but have not yet been published (Kendrick 1993, 1996).

There is one other group of organisms that has a "skeleton" consisting of many discrete plates and that has been the subject of a theoretical morphologic analysis. These are the receptaculitids, or "sunflower corals," which are actually extinct forms of dasyclad chlorophytes (multicellular green algae), and hence not animals at all. These interesting algae have a thallus, or skeleton, that is globose in shape in the adult form (and that, in many, superficially resembles an echinoid in profile). The thallus is composed of numerous small polygonal plates, or facets, that are arranged in circlets around a central nucleus. These plates are the distal ends of thin rods, or lateral shafts, which extend internally to the main axis of the alga.

The external arrangement of receptaculitid plates over the surface of the thallus forms a series of spirals radiating out from the nucleus, a series of spirals that superficially resembles the radiating series of spirals seen in the arrangement of seeds in sunflower heads. The respective number of dextral and sinistral (clockwise and counterclockwise) spirals in sunflower heads follows a ratio of sequential numbers in a Fibonacci series, a pattern that follows from any helical growth process in which new growth units are fit into the growing structure that has the most room or open space between previously created units (Stevens 1974). Gould and Katz (1975) pointed out, however, that the pattern of multiple radiating spirals seen in receptaculitids most often does not follow a pattern of Fibonacci fractions, and they were thus interested in the growth process by which the receptaculitid spiral forms were created. In particular, the tesselated facets in receptaculitid skeletons are developed in circlets about the nucleus rather than individually, and new spiral rows are added between the originals with growth. Gould and Katz developed a theoretical model of morphogenesis in receptaculitids that has five parameters:

1. the "nature of the surface" of the receptaculitid thallus, taken to be the radius of a sphere.

2. the "tightness of the spirals," where the spirals are modeled as logarithmic.

3, 4, 5. the "rate of increase in facet width," determined by an empirical function containing three constants.

One of the constraints, and insights, of their model is that the plates, or facets, of the receptaculitid thallus have maximum size limits, and this fact alone means that extra spiral rows must be intercalated between previously existing ones to maintain the tesselated pattern of plates in a continuous cover with the alga's growth. Facet width is thus a function of ontogeny, and this function is important to the timing of new plate-row intercalations. Rather than theoretically modeling this function, Gould and Katz (1975:11) stated that they "decided to specify the parameters of facet size empirically—i.e., we plotted plate width vs. circlet number for four spirals of the specimen . . . and fitted to these points an empirical equation." This equation sets facet width as a function of circlet number but contains three constants that are empirically determined by fitting the function to actual measurements from receptaculitids. Gould and Katz (1975:11) admit that they "seem to have violated a premise by using a complex equation with 3 parameters to represent the increase of facet width with circlet number. But this is a simple result of our decision to represent trends in facet size empirically." The "premise" seemingly being violated is the theoretical morphologic principle of producing a model of form that simulates morphologic complexity with the minimum number of parameters. In actuality, five parameters is not an inordinately large number. What is more a violation of the procedures of theoretical morphology is the use of an empirical function, with empirically determined constants, as part of the morphogenetic model. The fitting of generalized empirical functions to actual measurement data is a procedure of morphometrics, not of theoretical morphology, as is explained in chapter 1. However, Gould and Katz (1975) were successful in simulating a variety of actually existing receptatulitid morphologies using their model, in particular demonstrating how new plate rows must be intercalated to preserve full tesselation of the alga's surface.

As in the case of Raup's (1966) study of plate morphogenesis in echinoids, Gould and Katz (1975) were more interested in generating a theoretical morphologic model of receptaculitid facet-pattern development than in developing a theoretical morphospace of receptaculitid form. It would perhaps be possible, however, to modify their model to produce such a morphospace. Gould and Katz did sequentially modify the parameters of the model, even allowing the empirically deter-mined constants of the facet growth-rate function to vary, and found that variation in the "tightness of the spiral" affected the tesselation pattern most markedly and was most likely to generate hypothetical morphologies not seen in existent receptaculitids. Thus all the ele-ments of theoretical morphospace construction are present, and it would be very interesting to generate a theoretical morphospace of receptaculitid form and to explore the pattern of utilization—or non-utilization—of various regions of this morphospace by actual recepta-culitids.

## THEORETICAL MORPHOSPACES OF SEGMENTS AND SECTIONS?

What can I say? At the close of chapter 5, I noted that theoretical morphospaces of bivalved helicospiral form exist but simply (and sur-prisingly) have not yet been explored, but here I have no "theoretical morphospaces of segments and sections" at all to discuss. Yet skele-tons composed of segments and sections are a fundamental morpho-logic characteristic of the arthropods, and as mentioned at the begin-ning of the chapter, the arthropods are the dominant form of animal life on the planet (in terms of both sheer numbers and species di-versity).

The arthropods are also at the center of a vigorous debate among theorists concerned with deciphering the process of evolution of life on Earth, a scientific exchange in the literature and at professional meetings that has been christened by some paleobiologists as "The Great Cambrian Disparity Debate." We shall encounter this great de-bate in more detail in the next chapter. I close here simply with the observation that the arthropods are of tremendous importance in our understanding of the evolution of life on Earth, and a theoretical mor-phologic analysis of their myriad forms would be of equal importance to the scientific community. Creating a theoretical morphospace for arthropods will not be easy; it is in fact a monumental challenge. Will not some enterprising graduate student accept that challenge?

# The Time Dimension: Evolution and Theoretical Morphospaces

*The actual animals that have ever lived on Earth are a tiny subset of the theoretical animals that could exist. These real animals are the products of a very small number of evolutionary trajectories through genetic space . . . each perched in its own unique place in genetic hyperspace. Each real animal is surrounded by a little cluster of neighbors, most of whom have never existed, but a few of whom are its ancestors, its descendants and its cousins. Sitting somewhere in this huge mathematical space are humans and hyenas, amoebas and aardvarks, flatworms and squids, dodos and dinosaurs.*
—*Dawkins (1987:73, emphasis Dawkins's)*

In the previous four chapters, we examined theoretical morphospaces created for a variety of different growth systems and explored those theoretical morphospaces by means of the actual spectrum of form produced by organisms using those growth systems. Such an approach is static, however, and the evolution of life is dynamic. To understand what the evolution of life has both produced and not produced, we must compare form in theoretical morphospace. In the analogous "genetic hyperspace" envisioned by Dawkins (1987), humans, hyenas, dodos, and dinosaurs all occupy a region of that grand hypothetical morphospace. And we can learn a great deal about the evolution of life by understanding why humans are sitting over here in this region of the morphospace and the dinosaurs are sitting over there in that region of the morphospace and why there seem to be unoccupied regions of the morphospace where no one sits.

Humans and hyenas do exist at the present time, however, whereas dodos and dinosaurs are extinct (or, to be more precise, the great non-avian dinosaurs are extinct). We all occupy not only different spatial positions in the grand hypothetical morphospace but different positions in time as well. The techniques of theoretical morphospace analysis are particularly well suited to analyzing the evolution of morphology in time, and the techniques of empirical morphospace analysis are particularly unsatisfactory for that same task. The dimensions of a theoretical morphospace are mathematical constructs; they are timeless; and they exist in the absence of any organisms whatsoever. The dimensions of an empirical morphospace are determined by measurements taken from the very organisms themselves whose evolution through time is under study. That is, the dimensionality of an empirical morphospace is determined by morphology that exists either now or did in the past. An empirical morphospace can never reveal to us morphology that might have been or morphology that could be in the future, but theoretical morphospaces can.

What might we expect to see in the evolution of life in a theoretical morphospace? To explore that question, let us return to the adaptive, or fitness, landscape concept we covered in chapter 2. Kauffman (1993, 1995) conducted extensive computer simulations of evolution via the process of natural selection in $NK$ fitness landscapes, where $N$ is the number of genes under consideration and $K$ is the number of other genes that affect each of the $N$ genes. The fitness of any one of the $N$ genes is thus a function of its own state plus the states of the $K$ other genes that affect it, allowing us to model epistatic genetic interactions.

Kauffman (1993, 1995) demonstrated that two end-member landscapes exist in a spectrum of $NK$ fitness landscapes: a "Fujiyama" landscape at $K$ equal to zero, and a totally random landscape at $K$ equal to $N$ minus one, which is the maximum possible value of $K$. In the Fujiyama landscape, there is a single adaptive peak with a very high fitness value, with smooth slopes of fitness falling away from this single peak. Such a fitness landscape exists where there are no epistatic interactions between genes, where each gene is independent of all other genes. At the other extreme, every gene is affected by every other gene, and a totally random fitness landscape results, a landscape composed of numerous adaptive peaks, all with very low fitness values. A Fujiyama landscape has a single adaptive maximum. In a random landscape, any area in the landscape is just about the same as any other area. Between these two extremes is a spectrum of landscapes ranging from "smooth" (a few large peaks) to increasingly "rugged" (multiple smaller peaks) and from "nonisotropic" (landscapes where the large peaks tend to cluster together) to "isotropic" (landscapes where the large peaks are distributed randomly across the landscape).

Kauffman (1993, 1995) argued that the process of evolution on Earth appears to have taken place on rugged fitness landscapes and not on Fujiyama landscapes, smooth landscapes that he characterizes as the

183

Darwinian gradualist ideal. Computer simulations of the process of evolution via natural selection in rugged fitness landscapes reveals, on the one hand, that the rate of adaptive improvement slows exponentially as the evolving population climbs an adaptive peak but, on the other hand, that the highest peaks in the landscape can be climbed from the greatest number of regions! The latter conclusion is in accord with the empirical observation that convergent morphologic evolution has been extremely common in the evolution of life on Earth, and Thomas and Reif (1993) designated these peaks as "topological attractors."

If life evolved on rugged fitness landscapes, then epistatic interactions must be the norm, and the fitnesses of morphologic characters states must be correlated. Kauffman (1993, 1995) argued that the more interconnected the genes are, the more conflicting constraints will arise. These conflicting constraints produce the multipeaked nature of the rugged landscape. There is no single superb solution, as in a Fujiyama landscape. The conflicting constraints of the intercorrelated genes instead produce large numbers of compromise—less than optimum—solutions. A rugged landscape results, a landscape with numerous local peaks with lower altitudes.

Niklas (1997b) also contended this point, but from an engineering perspective. In relations optimization engineering theory, the number of equally efficient designs for a machine is proportional to the number of tasks that the machine must perform simultaneously. That is, when engineers design a machine to perform a single task, they commonly find only one configuration that can perform the task superbly well. When engineers design a machine that can perform numerous tasks simultaneously, they commonly find that many configurations can serve this purpose equally well. But even though the number of options has increased, each is a compromise configuration imposed by the conflicting design requirements of the multiple tasks, and the efficiency of performance of any particular task has decreased.

Niklas (1997b) also conducted computer simulations of evolution via natural selection by exploring a series of adaptive walks through Niklas and Kerchner's (1984) $p$–$\gamma$–$\phi$ theoretical morphospace, discussed in chapter 3. The hypothetical plants in the simulations have three tasks: light harvesting, mechanical stability, and reproduction. Plant morphologies that optimize each task are programmed to be selected for in the simulations. Seven possible simulation scenarios are possible in the $p$–$\gamma$–$\phi$ theoretical morphospace: three single-task adaptive walks, three dual-task adaptive walks, and one triple-task adaptive walk. Niklas (1997b) found that single-task simulations produced adaptive-walk pathways that wandered and meandered through the theoretical morphospace but seldom branched, whereas dual-task and triple-task simulations produced pathways that frequently branched (simulations of the dual-task walks and the triple-task walk are illustrated here in figure 7.1; note also the diversity of simulated plant morphologies resulting from the multitask pathways). Niklas concluded that the number of equally fit morphologies increases as the number of tasks that the hypothetical plants must perform simultaneously increases but that their maximum relative fitnesses decreases. Defining "complexity" as the number of biological tasks that an organism must perform simultaneously in order to survive, Niklas argued that the evolution of organic complexity may not impose severe limits on evolution but, rather, may actually make evolution easier by transforming a fitness landscape with a few high peaks into a fitness landscape with numerous low hummocks across which organisms may move more freely.

The peaks in an adaptive landscape also are not static. They may move their positions, and they may from time to time collapse and cease to exist (see Lewontin 1978). Strathmann (1978) suggested that "adaptive types" of organisms may have progressively vanished through the passage of geologic time because of the combined effects of occasional peak collapse and evolutionary specialization. Strathmann argued that organisms evolve from generalists with broad adaptations to specialists with narrow adaptations by a process of peak climbing in an adaptive landscape (he also has the organisms themselves steepening the slopes of the peaks, but this is a speculative complication not necessary for the development of the argument). The generalists, located at low-altitude positions in the landscape, can move about the landscape with little difficulty because from their relative perspective, the adaptive valleys are shallow. Specialists, however, are located in high-altitude positions at the very top of adaptive peaks and cannot budge from their high-fitness positions, because the adaptive valleys surrounding them are very deep. Under the expectations of the theory of natural selection, Strathmann (1978) contended that most organisms evolve over time from being generalists to being specialists. Late in the evolutionary scenario, most all organisms will be sitting on the tops of high-altitude peaks, surrounded by deep valleys. Now let us introduce an environmental catastrophe (such as an asteroid falling out of the sky) that results in the temporary destruction of two of the adaptive peaks, which leads to the extinction of the organisms inhabiting those peaks because the fitness of their areas of the landscape has dropped to zero. After the environment recovers and

185

FIGURE 7.1. Computer-simulated adaptive walks through theoretical plant morphospace. Each discrete step in the walk through the morphospace is represented by a bubble. The diameter of each bubble represents the volume of morphospace that had to be searched to locate the next most fit plant morphology. To the right of each morphospace diagram are examples of simulated plant morphologies residing on adaptive peaks for each walk. Figure 7.1A shows an adaptive walk that optimizes mechanical stability and reproductive success; 7.1B illustrates a walk optimizing light interception and mechanical stability; 7.1C depicts a walk optimizing light interception and reproductive success; and 7.1D shows a walk optimizing light interception, mechanical stability, and reproductive success. *Artwork courtesy of K. J. Niklas. Copyright © 1997 by the University of Chicago and used with permission.*

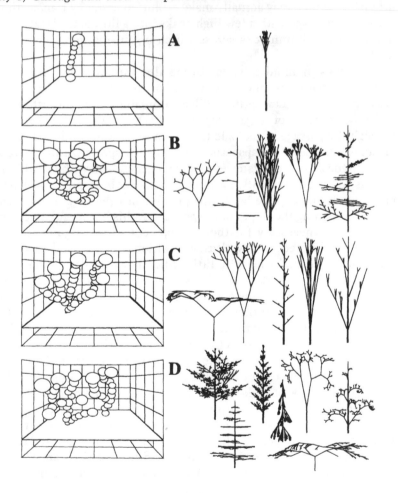

returns to the precatastrophe state, the two adaptive peak reappear but are not occupied, because their previous inhabitants are gone.

Now we enter the interesting conclusion of Strathmann's (1978) scenario. We have two vacant adaptive peaks on the landscape, two really nice pieces of real estate up for sale. Surely one or two of the neighboring organisms will evolve adaptations that will allow them to move onto these vacant adaptive peaks, no? Strathmann suggested that this may not be possible at this late and specialized stage in the organisms' evolution. The adaptive valleys that separate the existent organisms from the vacant peaks are simply too deep; the organisms have become too specialized to move very far on the adaptive landscape.

A startling conclusion of Strathmann's (1978) argument is that more adaptive-types of life may have existed in the earlier phases of evolution on the Earth than existing at present. Rather than steadily increasing the numbers of adaptive types of life, evolution may have led to the progressive loss of adaptive types through time. A similar idea, though formulated differently, underlies "The Great Cambrian Disparity Debate," a debate that has continued for a decade among theorists concerned with how evolution actually works. We shall encounter "The Great Cambrian Disparity Debate" again at the end of this chapter.

Strathmann's (1978) dead-end specialist scenario will result only if the positions of the adaptive peaks on the landscape do not move with time. The real adaptive landscape of life may be more mobile, with major shifting of the peaks across the landscape with time and organisms scrambling to keep up with them (Lewontin 1978). Do we have any evidence, one way or another, concerning the existence of stationary adaptive peaks with time? That question is particularly suited to theoretical morphospace analysis, as we shall see in actual examples in this chapter.

What about the startling conclusion of Strathmann's (1978) evolutionary scenario? Do we in fact see a progressive vacating of regions in theoretical morphospace with time? And if regions of theoretical morphospace do become vacant because the organisms that inhabited that space are extinct, does that space remain vacant from then on, as Strathmann suggested? There clearly has been a major vacation of theoretical morphospace by the nonbiconvex members of the phylum Brachiopoda. The nonbiconvex regions of bivalved theoretical morphospace, once inhabited by numerous strophomenidine and productidine brachiopods (figure 5.14), became vacant with the extinction of those higher taxa, and those regions of morphospace have remained vacant

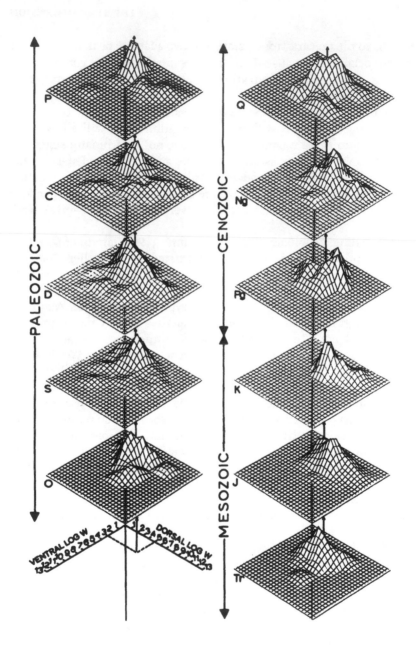

FIGURE 7.2. The actual pattern of morphologic evolution exhibited by the biconvex Brachiopoda for the past 500 Ma, as seen in the bivalved theoretical morphospace given in figure 5.1. The vertical dimension in each time-horizon diagram is morphologic frequency. The position of the composite peak of the biconvex brachiopod frequency distribution (see figure 5.10) is given with respect to the axes of the morphospace in the graph in the lower left corner of the figure and is indicated by the vertical arrow in each time-horizon morphologic frequency diagram. The diagonal line in the lower left corner graph, indicating the location of equibiconvex shells in the morphospace, is also reproduced in each time-horizon diagram. Time-horizon abbreviations: O, Ordovician; S, Silurian; D, Devonian; C, Carboniferous; P, Permian; Tr, Triassic; J, Jurassic; K, Cretaceous; Pg, Paleogene; Ng, Neogene; Q, Quaternary.

ever since. In chapter 6, we encountered vacant regions in Reif's (1980) "elasmobranch dermal skeleton" theoretical morphospace. That region of theoretical morphospace remains vacant today, but Reif (1980:356) believes that "if more data were available on Paleozoic shark squamations, this area could probably be filled with real types." Reif believes that although this region was once occupied, it is now vacant. Because neither Reif (1980) nor McGhee (figure 5.14) offers quantitative data from extinct Paleozoic sharks or nonbiconvex brachiopods, the absolute magnitude of this vacating of theoretical morphospace is currently not known.

Has the evolution of anything in theoretical morphospace been rigorously quantified? The answer is yes, but the number of such studies at present are few. An introductory example is given in figure 7.2, which shows the pattern of morphologic evolution of the biconvex brachiopods for the past 500 Ma, as seen in the bivalved theoretical morphospace discussed in chapter 5 (see figure 5.1). When I first showed this figure to the functional morphologist Steve Stanley, his reaction was immediate: "It looks like a cat playing under a blanket." Note, however, that even as the cat alternately attacks imaginary mice to the left and imaginary Martians to right, the cat itself stays pretty much in the same position under the blanket. I much prefer Steve's cat to poor Schrödinger's and shall return to figure 7.2 later.

## MASS EXTINCTION IN THEORETICAL MORPHOSPACE

*The ecological or evolutionary impact of an extinction event resides not simply in the number of taxa lost, but in the loss of what could be termed (with some trepidation) biodisparity, the range of morphologies or other*

189

*attributes within a clade. . . . A variety of methods are available for quantifying biodisparity, ranging from ecomorphological studies . . . to the construction of theoretical or empirical morphospaces used to track clades through geologic time (reviewed by McGhee 1991; Foote 1991, 1993). As Foote (1992) points out, if extinction is effectively random with respect to morphology, a disproportionately large number of extinctions is required to reduce morphological variety substantially. On the other hand, extinctions might be selective in morphospace just as they are taxonomically. This has not been explored, however, and we do not know if taxa lying in different regions or densities of morphospace vary predictably in risk.*

*Jablonski (1994:14)*

The ultimate measure of biotic crises in the history of life is the loss of species diversity from one geologic time interval to another, and thus most studies of mass extinction have focused on diversity changes through time (McGhee 1989). The relationship between taxonomic diversity and morphologic diversity is not a simple one-to-one linear function. Large losses in taxonomic diversity may merely "thin" the number of species present without having much effect on the total range of morphologies present. Alternatively, taxonomic diversity losses may preferentially occur in species possessing particular morphologies, producing a postextinction range of morphology that is very different from that present before the extinction event. Theoretical morphospaces are particularly suited to the study of the relationship between taxonomic and morphologic diversity changes through time, as the dimensions of a theoretical morphospace are sample independent, which is not the case for empirical morphospaces. The dimensionality of a theoretical morphospace does not change, regardless of whether a large number of species are plotted in that hyperspace (preextinction time) or a small number of species are plotted (postextinction time).

An early study that considered the effect of mass extinction in theoretical morphospace is that by Ward (1980), which we encountered in chapter 4. Ward (1980) actually was interested in the biologic differences between ammonites and nautilids and in the different regions these two groups of animals occupy in theoretical morphospace (see figure 4.11), and not primarily in the effect of mass extinction. But because the ammonites did not survive the end-Cretaceous mass extinction, and the nautilids did, his study automatically included the effect of that extinction event. Obviously, as the ammonites became extinct in the end-Cretaceous event, the region of morphospace formerly occupied by these animals was vacant in the early Tertiary. Of

**190**

more interest is the range of morphologies seen in the surviving ecto-cochleate cephalopods, the nautilids.

Ward (1980) found a very interesting shift in the frequency distribu-tion of nautilid morphologies in $W–D–T–S$ theoretical morphospace following the end-Cretaceous mass extinction (figure 7.3). Values of $D$ remained about the same, but many more nautilids evolved with lower values of $S$ in the Tertiary than were present in the Cretaceous or Ju-rassic, as can be seen in the spacing of the contours in figure 7.3. The peak, or mode, of the morphologic frequency distribution also exhibits this shift, moving from a position of $S$ equal to about 1.15 in the Cretaceous to a position of $S$ around 0.9 in the Tertiary (figure 7.3). Ward also documented a similar decrease in values of $W$ that occurred in nautilid shell form from the Cretaceous to the Tertiary. What does this mean? "Tertiary nautilids had shells which, on the average, were more compressed and with lower whorl expansion than were evolved by Mesozoic nautilids" (Ward 1980:38); in fact, these Tertiary nautilids shifted into the region of morphospace that had previously been occu-pied by the ammonites!

Ward (1980:32) does not think that this pattern is a coincidence and

FIGURE 7.3. Frequency distribution of nautilid shell morphologies in the Jurassic, Cretaceous, and Tertiary, in the $D–S$ plane of Raup's $W–D–T–S$ theoretical morphospace (1966). Contours measure the increase in density of taxa per unit area of the plot, and the cross marks indicate the peak morphologic position. *From Ward 1980. Reprinted from* Paleobiology *and used with permission.*

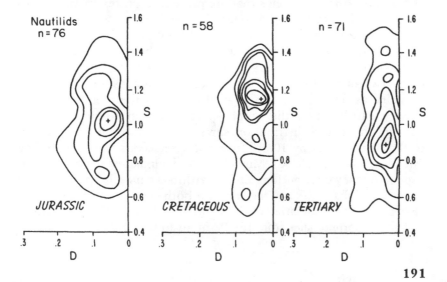

suggests that "the terminal Cretaceous extinction of ammonites may have opened up new opportunities for nautilid evolution during the Tertiary, because Tertiary nautilids are dominated by moderately compressed, hydrodynamically efficient shell shapes which were rarely present among Jurassic and Cretaceous nautilids, but common among ammonites." He does note, however, that the region of ammonite morphospace that was successfully invaded by nautilids is very small compared with the former total range of form exhibited by the ammonites. The great majority of theoretical morphospace once occupied by the ammonites and their kin (see figure 4.9) was vacated when the ammonoids became extinct and remains vacant today.

The exploration of the morphologic effects of mass extinction in theoretical morphospace was the explicit goal of McGhee's studies (1995a,b; 1999a), although with brachiopods rather than cephalopods. Interestingly, McGhee (1995a,b) found that the morphologic response of biconvex brachiopods to the end-Permian and end-Cretaceous mass extinctions is very similar, even though the phylogenetic lineages in these two extinction episodes are very different and even though these episodes are separated by some 180 Ma of geologic time. From the total Phanerozoic pattern for the biconvex brachiopods given in figure 7.2, the morphologic frequency distribution in the theoretical morphospace for the Permian is extracted in figure 7.4A, the Triassic in figure 7.4B, the Cretaceous in figure 7.4C, and the Paleogene in figure 7.4D. Thus figure 7.4 illustrates the frequency distribution of biconvex brachiopod morphologies present in preextinction times in the left half of the figure (figure 7.4A and C), and the frequency distribution of brachiopod morphologies that are present in postextinction times in the right half of the figure (figure 7.4B and D).

In comparing the left half of figure 7.4 with the right half, we can see that the morphologic response of biconvex brachiopods to the end-Permian and end-Cretaceous mass extinctions was quite similar. In both cases, during the mass extinction there was a loss of morphologic diversity, as represented by the areal extent of the morphologic frequency distribution in the morphospace. The areal extent of the morphologic frequency distribution shrank by 8.7% from the Permian to the Triassic and by 10.2% from the Cretaceous to the Paleogene (McGhee 1995a,b; 1999a). Moreover, the boundaries of the morphologic frequency distributions shifted within the morphospace, and the direction of that shift was the same for both extinction episodes. The changes in the boundaries of the frequency distributions of brachiopod shell morphologies from the Permian to the Triassic are summarized in figure 7.5A and from the Cretaceous to the Paleogene in figure 7.5B.

From the Permian to Triassic (figure 7.5A), there was a major loss of brachiopods having highly inequivalved shells with very flat dorsal valves (high dorsal $W$ values). But new brachiopod morphologies appeared in the Triassic with flatter ventral valves (higher ventral $W$ values), so that the median ventral valve position shifted from a value of

**FIGURE 7.4.** Frequency distribution of biconvex brachiopod shell morphologies in the (A) Permian, (B) Triassic, (C) Cretaceous, and (D) Paleogene, in the theoretical morphospace given in figure 5.1. Contours measure the increase in density of genera per unit area of the plot. Taxa that fall along the diagonal line in the plots are equibiconvex. *Modified from McGhee 1995a.*

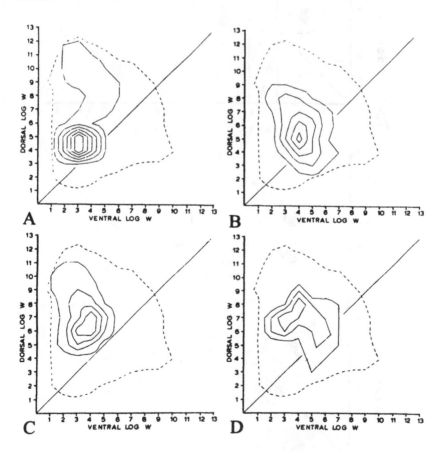

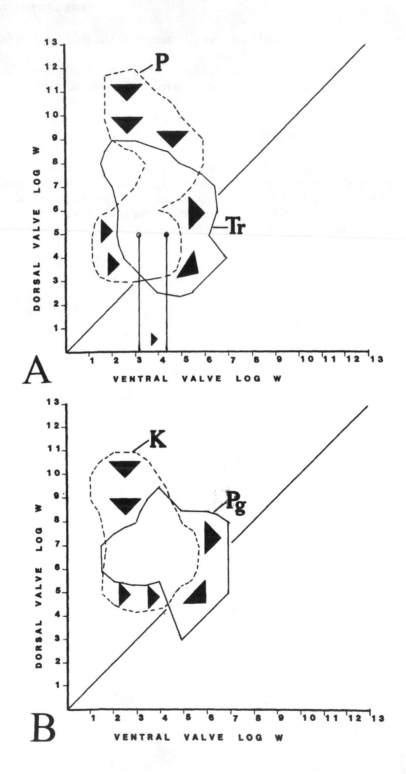

the logarithm of $W$ of 3.16 to a value of 4.34 (figure 7.5A), which is one of the few statistically significant shifts (at the 1% level) in median brachiopod shell form that occurred in geologic time (McGhee 1995a,b; 1999a). The end result of the mass extinction was that brachiopods retreated from dorsal high $W$ regions of the morphospace but spread into ventral higher $W$ regions. That is, the entire frequency distribution of brachiopod shell forms shifted down and to the right in the morphospace and toward the diagonal line of equiconvexity, where shell forms with ventral and dorsal valves of equal $W$ values are located. In comparing figures 7.4A and B and 7.5A, we can see that the Triassic biconvex brachiopod frequency distribution is much more centered about this diagonal line in the morphospace than the Permian; that is, postextinction Triassic biconvex brachiopod shells are more equibiconvex and have lower area-volume ratios than do preextinction Permian ones.

The nature of the shift in brachiopod morphologic frequency distributions from the Mesozoic to Cenozoic (and through the end-Cretaceous mass extinction) and Paleozoic to the Mesozoic (through the end-Permian mass extinction) is very similar (cf. figure 7.5A and B), although the shift in median position of the ventral valve toward higher $W$ values is statistically only "weakly significant" (at the 5% level) in the transition from the Cretaceous to the Paleogene (McGhee 1995a,b; 1999a). In both transitions, the boundaries of the brachiopod morphologic frequency distribution shifted downward (toward more convex dorsal valves) and to the right (toward flatter ventral valves), so that the boundaries of the resultant distribution are more centered about the diagonal line in the geometric continuum (figure 7.5A and B). Early Cenozoic biconvex brachiopods, like early Mesozoic brachiopods, have shells that are more equibiconvex and have lower area-volume ratios than those of the previous era.

From a functional, rather than a purely geometric, perspective, it appears that biconvex brachiopods in their evolutionary history have repeatedly evolved less than optimum shell forms (see chapter 5 for a discussion of optimum brachiopod shell form). That is, periods of brachiopod morphologic diversification are characterized by expansion

FIGURE 7.5. Directional change in the boundaries of biconvex brachiopod morphological frequency distributions from the Permian to the Triassic (A) and from the Cretaceous to the Paleogene (B). Vertical lines along the x-axis in figure 7.5A depict the statistically significant shift of the median value of $W$ in ventral valves toward higher values that occurs in this interval of time. *Modified from McGhee 1995a.*

of the boundaries of the morphologic frequency distribution into regions of the theoretical morphospace where shell forms having higher area-volume ratios than the median position are found, particularly into the dorsal-valve high $W$ regions of the morphospace (see figures 5.1 and 5.10 for illustrations) where inequivalve ventri-biconvex to plano-convex morphologies occur. The mass extinction episodes at the end of the Paleozoic and Mesozoic are evidenced by retreats and contractions of the boundaries of the morphologic frequency distribution back to regions of the theoretical morphospace characterized by shell forms with lower area-volume ratios and more equibiconvex, spherical morphologies. Thus the biconvex brachiopods have both vacated and reinvaded regions of theoretical morphospace with time (see figure 7.2).

## EVOLUTIONARY TRENDS IN THEORETICAL MORPHOSPACE

Dave Raup once related a story to me about people's early reaction to his study of the distribution of ammonoid form in theoretical morphospace (see figure 4.9 of this book). Most everyone wanted to know what the distribution of ammonoid form looked like in time, that is, how the frequency distribution moved around in the theoretical morphospace from geologic period to geologic period. It is obvious that the ammonoids evolved, so they had to change position in morphospace, right? Dave told me of their disappointment, over and over again, when he told them that as far as he could see (with the 405 species in his sample), the frequency distribution remained in the same position with time. In the overall ammonoid morphologic distribution, differing taxonomic groups occupied different regions of morphospace, and they seemed to stay in the same place with time as well. Ammonoid species went extinct, and new ammonoid species evolved, but all this evolutionary change seemed to be simply repeating the same morphologic themes over and over again, with no major directional shifts in $W$–$D$–$T$–$S$ theoretical morphospace. It was as if a "regional morphologic stasis" zone existed in theoretical morphospace for the ammonoids.

Do within-clade evolutionary trend show up in theoretical morphospace? Because Raup's (1967) early study concentrated on ammonoids, he missed the interesting between-clade shifts seen in theoretical morphospace by Ward (1980) for ammonites and nautilids before and after the end-Cretaceous mass extinction, as noted in the previous section. And also in the previous section we saw that certain major evolutionary events (such as mass extinctions) do show up in in within-clade morphologic shifts in theoretical morphospace, or at least they do for the clade of biconvex brachiopods.

Other studies suggest that Raup's early observation of a regional morphologic stasis in theoretical morphospace might indeed be true for the clade of the ammonoids. A series of more recent combination studies (using both theoretical and empirical morphospaces) examined morphologic patterns of evolution in the ammonoids (Saunders and Swan 1984, Saunders and Work 1996, Dommergues, Laurin, and Meister 1996). Figure 7.6 gives a morphologic frequency distribution for Late Carboniferous ammonoids in the $W–D$ and $S–D$ sections of the $W–D–T–S$ theoretical morphospace (characteristic morphologies found in this region of the theoretical morphospace are given in figure 7.7). The morphologic frequency distribution of shell forms in these Late Carboniferous ammonoids in the $W–D$ plot (figure 7.6) is virtually identical to that seen for ammonoids throughout their existence in geologic time (figure 4.9). The single broad peak seen in Raup's (1967) study appears to consist of a series of smaller "peaklets" in the Carboniferous (cf. figures 4.9 and 7.6), but this could simply be due to differing algorithms used in contouring the data. The major orientation and spread of the distribution in theoretical morphospace is the same.

Such a regional morphologic stasis in theoretical morphospace suggests the existence of a stationary adaptive peak through geologic time, an evolutionary stable strategy that worked well for the ammonoids, at least up to the point that they were eliminated in the end-Cretaceous mass extinction. Functional morphologic arguments for the existence of such an adaptive peak were given in chapter 4.

Within-clade regional shifts in theoretical morphospace for the ammonoids, however, are seen in the Early Jurassic (figure 7.8). Dommergues, Laurin, and Meister (1996) documented that Early Jurassic ammonoids are in slightly different region of the $W–D–T–S$ theoretical morphospace than are Late Carboniferous ones. In general, Early Jurassic ammonoids have higher values of $W$ and $D$, and lower values of $S$, than do Late Carboniferous ammonoids. That is, Early Jurassic ammonoids tend to be more compressed and evolute than Late Carboniferous ones (cf. figures 7.7 and 7.8), and the distribution of shell forms is shifted more toward the boundary between the whorl-overlap and non-whorl-overlap regions of the theoretical morphospace (cf. figures 7.8 and 4.9).

The reason for these regional differences in theoretical morphospace occupation by Early Jurassic and Late Carboniferous ammonoids

197

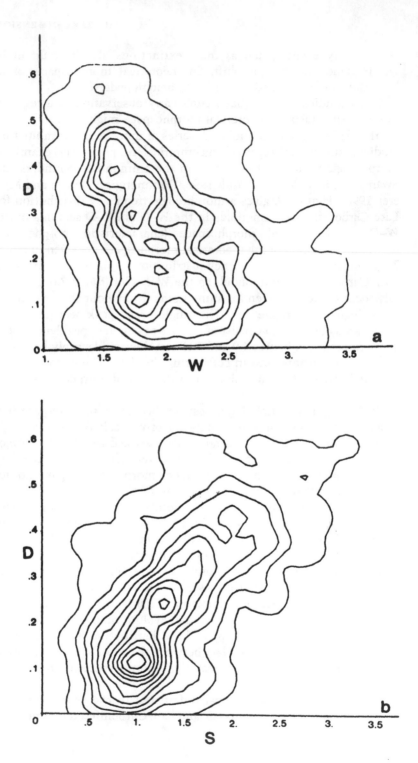

remains unclear. Dommergues, Laurin, and Meister (1996) noted that Early Jurassic ammonoids had just survived the end-Triassic mass extinction and its consequent ""bottleneck" of species diversity and that thereafter the surviving ammonoids were evolving in a global marine-transgressive context that may have offered weaker ecological constraints than those experienced by Late Carboniferous ammonoids. Thus mass extinction may once again be the trigger for major shifts in theoretical morphospace, as seen for biconvex brachiopods.

More important, figure 7.8 allows me to explain a major problem of empirical morphospaces. The theoretical parameters $W$, $D$, and $S$ are geometric and exist in the absence of any measurements taken from actual ammonoids. The dimensions of the $W$–$D$–$T$–$S$ theoretical morphospace remain the same, whether Late Carboniferous ammonoids are plotted in the morphospace (figure 7.6) or Early Jurassic ammonoids are plotted or *both* are plotted together (figure 7.8) in a direct contrast of morphologies from those two time horizons.

This measurement-indcpendent dimensional property is not shared by the empirical morphospaces given in Saunders and Swan (1984) and Dommergues, Laurin, and Meister (1996). For example, Dommergues, Laurin, and Meister (1996:235–236) noted that their empirical morphospace analysis reveals Early Jurassic ammonoid form as a "homogeneous distribution with no significant gap, trend, or strong aggregates," but they found that the empirical morphospace analysis of Saunders and Swan (1984) shows a Late Carboniferous ammonoid morphologic distribution "that is both more disjointed and locally strongly aggregated." Are these differences real, or are they artifactual? In fact, the two empirical morphospaces cannot be directly compared, as Dommergues, Laurin, and Meister (1996:236) admitted that "the characters used by those authors are not the same as in the present study."

Measurements taken from actual morphologic characters of actual

---

FIGURE 7.6. Frequency distribution of Late Carboniferous ammonoid shell morphologies plotted (a, upper figure) in a $W$–$D$ section of the $W$–$D$–$T$–$S$ theoretical morphospace (see figure 4.8), and (b, lower figure) in a $S$–$D$ section of Raup's $W$–$D$–$T$–$S$ morphospace (1966). Contours measure the increase in density of taxa per unit area of the plot. Note that the frequency distribution of Late Carboniferous ammonoids in the $W$–$D$ space approximates the distribution shown for ammonoids as a whole throughout geologic time, given in figure 4.9 (rotate figure 4.9 counterclockwise 90° in the comparison, so that the $W$–$D$ axes have the same orientation). *From Saunders and Swan 1984. Reprinted from* Paleobiology *and used with permission.*

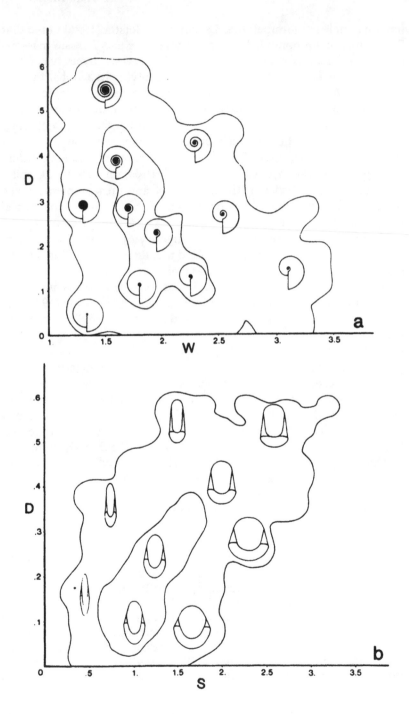

ammonoids determine the dimensions of the empirical morphospaces of both Saunders and Swan (1984) and Dommergues, Laurin, and Meister (1996). Yet because the characters measured differ in each study, the empirical morphospaces produced *possess different dimensions* in each study, and points within those empirical morphospaces cannot be directly superimposed on a single time-comparison plot. Note that this is not a problem with theoretical morphospace analyses, as such a single time-comparison plot of Late Carboniferous versus Early Jurassic ammonoid morphologic distributions is not only possible with theoretical morphospaces—the reader can see an actual one in figure 7.8.

Finer-scale (at the zonal level) evolutionary trends in Middle Jurassic ammonites also were documented in theoretical morphospace by Bayer and McGhee (1984; 1985a,b); and McGhee, Bayer, and Seilacher (1991). Ammonites in the southern German epicontinental sea during the Middle Jurassic repeatedly evolved similar morphologies in several different lineages, a phenomenon known as "iterative evolution" or "iterative morphologic series." Ammonite species at the beginning of each iterative morphologic series possessed shells that were inflated (having high $S$ values) and evolute (having high $D$ values). These species are successively replaced by species with shells that are progressively more compressed and discoidal (having low $S$ values) and that are progressively more involute (having low $D$ values) with time. In terms of a $D$–$S$ section of the $W$–$D$–$T$–$S$ theoretical morphospace, southern German Middle Jurassic ammonites repeatedly track from the upper right corner of the theoretical morphospace (figure 7.9) to the lower left corner. This phenomenon is exhibited by no fewer than four separate higher taxa of ammonites (figure 7.9). Each of the iterative morphologic series appears to have been triggered by repetitive transgressive-regressive cycles of sea level, which repeated these sequential changes in marine environments in the southern German epicontinental sea (Bayer and McGhee 1985a; McGhee, Bayer, and Seilacher 1991). Thus these studies suggest that a more mobile adaptive landscape exists at smaller scales in the overall ammonite morphologic frequency distribution.

Larger-scale regional shifts in theoretical morphospace are exhibited by the biconvex brachiopods, although the nature of these regional

---

FIGURE 7.7. An illustration of the range of existent shell geometries found in Late Carboniferous ammonoids, whose frequency distributions were given in figure 7.6. *From Saunders and Swan 1984. Reprinted from* Paleobiology *and used with permission.*

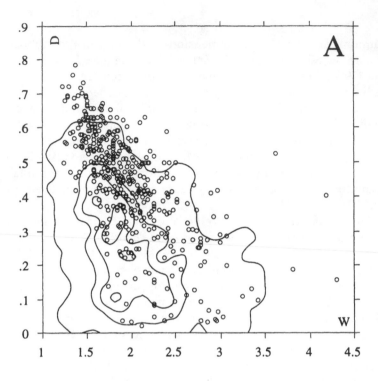

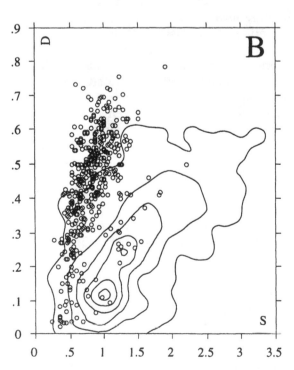

shifts does not argue against the concept of an evolutionary stable strategy or a stationary adaptive peak through geologic time for the biconvex brachiopods (see figure 7.2). The early evolutionary radiation of biconvex brachiopods can clearly be seen in theoretical morphospace (figure 7.10). Although the boundaries, or perimeters, of the morphologic frequency distributions of brachiopod shell forms change through time (figures 7.10 and 7.2), the position of the peak, or mode, of the frequency distribution does not. Thus the majority of brachiopods (even in the Devonian) are still found in the lower left corner of the theoretical morphospace, which is the region of spherical shell forms having low area-volume ratios and the optimum region for these filter-feeding organisms (as discussed in chapter 5). Only during periods of mass extinction are there significant shifts in the peaks of the morphologic frequency distributions for brachiopods, as discussed earlier in this chapter.

What can be seen to change is the progressive evolution of species through time into regions of the theoretical morphospace that are farther from the optimum region, as exhibited by the progressive expansion of the boundary of the brachiopod morphologic frequency distribution from the Ordovician to the Devonian (figure 7.10A, B, and C), an expansion summarized in theoretical morphospace in figure 7.10D. The Devonian was the "golden age" for the biconvex brachiopods, and their evolutionary expansion into less than optimal regions of theoretical morphospace was subsequently and repeatedly curtailed by four successive mass extinctions in geologic time (McGhee 1995a).

Other studies of the evolution of shell form in theoretical morphospace include those of Davoli and Russo (1974) and Cain (1977) with gastropods. Davoli and Russo (1974) explored evolutionary changes in five species of the gastropod *Subula* from the Miocene to the recent in a $W$-$T$ slice of Raup's (1966) $W$-$D$-$T$-$S$ theoretical morphospace. But in his analysis of gastropod form, Cain abandoned the $W$-$D$-$T$-$S$ theoretical morphospace in favor of an alternative "spire index" theoretical

FIGURE 7.8. Distribution of Early Jurassic ammonoids, shown by individual points (open and filled circles), and Late Carboniferous ammonoids, shown by contour lines (given in figure 7.6), in the $W$-$D$ plane (figure 7.8A) and the $D$-$S$ plane (figure 7.8B) of Raup's $W$-$D$-$T$-$S$ theoretical morphospace (1966). The Early Jurassic ammonid distribution is shifted to the higher $W$ and $D$ regions and the lower $S$ regions, in the theoretical morphospace relative to the distribution of Late Carboniferous ammonoids. *From Dommergues, Laurin, and Meister 1996. Artwork courtesy of J.-L. Dommergues. Reprinted from* Paleobiology *and used with permission.*

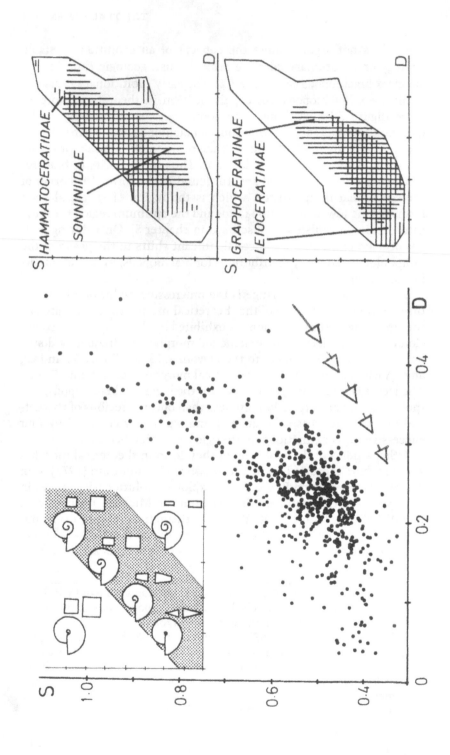

morphospace of conical form (as pointed out in chapter 4). In subsequent examinations of gastropods through geologic time, Cain found that archaeogastropods were much more widely distributed in the theoretical morphospace of conical form in the Paleozoic than they were in the Mesozoic and Cenozoic and that the retreat of archaeogastropods from regions of the theoretical morphospace was matched by the occupation of those same regions by caenogastropods. Invoking the theory of natural selection and the argument that shell form corresponds to specific habitat (see the discussion in chapter 4), Cain argued that the observed regional displacement of the archaeogastropods in theoretical morphospace reflects the active competitive exclusion, by caenogastropods in the Mesozoic and Cenozoic, of archaeogastropods from habitats they used to occupy. Thus Cain may provide us with an example of the vacation of a region of theoretical morphospace by one group of organisms, owing to the active moving into that same region by others, a scenario not found in Strathmann's (1978) model.

Theoretical morphospaces of branching form in both animals and plants were discussed in chapter 3. The use of those theoretical morphospaces to explore the time dimension has been limited, but there is no good reason for this trend to continue. Such studies are clearly possible; for example, Cheetham and Hayek (1983) themselves examined the evolution of 17 species of bryozoans in geologic time in their $\beta$–$\tau$-$R_g$–$R_l$ theoretical morphospace (covered in chapter 3). They found a decrease in the variability of branch dimensions in adeoniform cheilostomes from Paleocene to Recent and suggested that this could reflect the selective advantage of adhering more closely to regular growth, as maximizing growth regularity maximizes the colony's ability to utilize interbranch space to increase the colony's surface area. Although dealing with marine animals, Cheetham and Hayek (1983: 259) noted a similar evolutionary observation for terrestrial plants, in

FIGURE 7.9. Distribution of some Middle Jurassic ammonite faunas in the $D$–$S$ plane of Raup's $W$–$D$–$T$–$S$ theoretical morphospace (1966). The ammonite faunas repeatedly begin with shell geometries with higher values of $S$ and $D$, which are then repeatedly replaced with later species having lower values of $S$ and $D$ (shown by the direction of the arrows), a phenomenon known as "iterative evolution." The entire data set is given in the cloud of points on the left, and the subdivision of this cloud into major taxonomic groups is given in the two smaller graphs on the right. An illustration of the characteristic shell geometries to be found in this region of the theoretical morphospace is shown in the inset graph in the upper left corner of the figure. *Modified from Bayer and McGhee 1984.*

205

**FIGURE 7.10.** The early evolutionary radiation of biconvex brachiopods in the theoretical morphospace given in figure 5.1. Illustrated are the frequency distributions of biconvex brachiopod shell morphologies in the (A) Ordovician, (B) Silurian, and (C) Devonian, and a summary of directional change in the boundaries of the morphological frequency distributions is given in (D). Contours measure the increase in density of genera per unit area of the plot. *Modified from McGhee 1995a.*

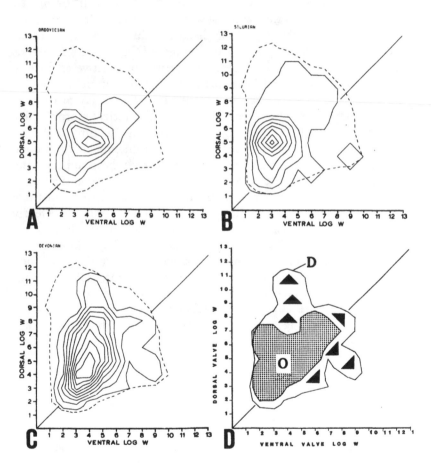

that "a decreasing variation in branching in the evolution of early land plants has been interpreted as evidence for the development of regular patterns from initially random branching (Niklas 1982)." Cheetham and Hayek thus pointed the way for future evolutionary studies of bryozoan form in theoretical morphospace, having both created a theoretical morphospace and demonstrated its usability. Plotting 17 species in theoretical morphospace is just a beginning, however, and the further documentation of the actual pattern of bryozoan evolution through time is a Ph.D. dissertation topic waiting to be seized.

Niklas (1982, 1986, 1994, 1997a,b), whose work with plant evolution was cited by Cheetham and Hayek (1983) in the previous paragraph, also examined the evolution of branching form in theoretical morphospace, although his exploration used a computer simulation of evolutionary process (see figures 3.9 and 7.1) rather than actually exploring theoretical morphospace by plotting the distribution of existent form through geologic time. Niklas (1997a,b) outlined interesting general conclusions from his simulation of evolution studies, some of which were discussed at the beginning of this chapter, yet it remains for some enterprising young researcher or graduate student to actually document the unfolding of plant evolution through time in theoretical morphospace.

## THE "CAMBRIAN EXPLOSION" IN THEORETICAL MORPHOSPACE?

*Several workers (e.g., Gould 1991; Hickman 1993a,b) have pleaded for the use of theoretical morphospaces capable of accommodating the range of conceivable, not just realized, forms. Unfortunately, such systems are more easily desired than obtained.*

*Foote (1995:295–296)*

Nowhere is the chasm between the desire for a theoretical morphospace and the obtaining of such a morphospace more apparent than in the case of the Cambrian arthropods or, to be even more specific, for the Burgess Shale arthropods. A contentious debate (dubbed by one morphologist as "The Great Cambrian Disparity Debate") concerning these peculiar arthropods (and other organisms, as the debate spread) began in paleobiological circles at the beginning of this decade and continues today (to list only a few of the principal works: Gould 1989, 1991, 1993; Briggs and Fortey 1989; Fortey 1989; Bard 1990; Foote and Gould 1992; Briggs, Fortey, and Wills 1992; Hickman 1993a; McShea 1993; Ridley 1993; Wills, Briggs, and Fortey 1994; Kendrick 1996; Wills 1996). The focus of the debate, the vortex about which the winds of the storm have twisted, is a question that at first glance seems quite simple: was the disparity in anatomical design greatest early in the history of animal life? Or more specifically, was the biodisparity of the Burgess Shale arthropod fauna much greater than that seen in modern surviving arthropods? Or more generally, are the branches in the tree of life arranged in the shape of a cone, with ever more branches (and hence diversity of life) increasing and expanding from the distant past to the present, or is the tree of life like a bush that someone has progressively pruned, so that many branches existed at its base in the Cambrian but only a few have managed to extend upward in time to the present?

Unfortunately, there is no theoretical morphospace for arthropod form (as lamented in chapter 6). Arthropods are morphologically very complex organisms. In the absence of a theoretical morphospace, attempts have been made to address "The Great Cambrian Disparity Debate" by using empirical morphospaces. I contend that the debate will never be—indeed *can* never be—conclusively closed by the usage of empirical morphospace analysis. In discussions with colleagues, I have pointed out that trying to analyze the "Cambrian Explosion" with empirical morphospace techniques is like trying to analyze the evolution of the universe from the "Big Bang" to the present *from the inside*. An analogy that usually brings me puzzled expressions, because how else can the evolution of the universe be analyzed? We are positioned firmly in the universe, a universe that continues to expand around us, and thus we are embedded in a universe whose spatial distances are constantly changing (at least in geologic time scales). All our measurements of distances to far stars and galaxies will be wrong in only a few hundred millions of years, as the very dimensions of the universe itself will have changed. We will be in the same "from the inside" position in analyzing the question of the expansion or non-expansion of arthropod morphologic distances from the "Cambrian Explosion" to the recent if we use measurements taken from those very same arthropods *to define the dimensions of the space in which we are measuring their morphologic distances.*

Imagine now that we could somehow step outside the universe and examine the expansion and evolution of the universe from a frame of reference that is not changing. And that is what we could do if we could analyze the "Cambrian Explosion" using theoretical morphospace techniques. The challenge is clear, and it needs someone to accept it.

# VIII

# Theoretical Models of Morphogenesis: An Example

*In my own experience with laboratory exercises in which students explore simple computer models of shell form, those students who become most excited are invariably motivated by an interest in improving the models.*
*—Hickman (1993b:171)*

Growth is the subject of this chapter, not morphospace. We now begin our consideration of models of morphogenesis, the second research focus of theoretical morphology, and leave our previous considerations of theoretical morphospaces. Many readers may ask, how does one actually model form or growth? How does one begin? How does one proceed? Those are good questions, and the purpose of this chapter is to give a simple but concrete example of theoretical morphogenetic modeling. In chapter 9 we shall examine in detail the various morphogenetic models that have been created to analyze growth in organisms with accretionary growth systems. In chapter 10 we shall examine morphogenetic models of branching and discrete growth systems, as well as series models, physicochemical developmental models, and even some theoretical morphogenetic models of behavior itself.

For the example of theoretical morphologic modeling outlined in this chapter, I have chosen accretionary growth systems, which, as I mentioned in chapter 4, are some of the easiest to simulate. The product of an accretionary growth system most familiar to us is the shells of invertebrate animals. Mathematical models of shell form historically have been of two general types: they attempt either to simulate existing form or to model the basic growth system that produces form

(McGhee 1981). The early "conchyliometrists" were primarily interested in the former, and much debate and disputation was generated
over the "goodness of fit" of various models to actual existent shells,
on abstract geometric grounds (see, e.g., Moseley 1838, 1842, and Naumann 1845). The opposite approach is often taken in theoretical morphology, however, in which geometric models are used to create new
and hypothetical morphologies, rather than simply seeking to characterize existent forms by an abstract model. Likewise, with models of
morphogenesis, one can develop a model that simulates the outcome
or result of the growth process, thereby assuming that the actual
growth process is similar in degree of complexity to the model, or one
can simulate the growth process itself.

   This is a fun chapter, a stroll through some of the mathematical
gardens of theoretical morphology. There is no examination at the end,
nothing to worry about. If you are not in the mood for mathematics
with your coffee this morning, do feel free to skip to the more general
discussion of morphogenetic models in chapters 9 and 10. But please
come back to this chapter and read it later; otherwise you will miss
out on some of the fun of theoretical morphology.

## TO BEGIN: SIMULATING SHELL FORM WITH
## A LOGARITHMIC SPIRAL IN A PLANE

Early naturalists were impressed—as we still are today—with the
ubiquity of the spiral form in nature (Stevens 1974). The peculiar geometric properties of the logarithmic spiral (see figure 8.1A), in particular, drew much attention from the early conchologists (see Thompson
1917, 1942). The great morphologist D'Arcy Thompson devotes more
than 100 pages of text to considering the many mathematical properties of the "equiangular" or logarithmic spiral and the many organic
forms in which such spirals are found (Thompson 1942:748–849).

   In essence, the logarithmic spiral is the product of simple exponential growth, that is, of an equation of the form

$$r_\theta = r_0 e^{c\theta} \tag{8.1}$$

where the radius of the spiral $(r_\theta)$ at any given angle of rotation $(\theta)$ is
equal to the initial radius $(r_0)$ multiplied by the natural base of logarithmic functions $(e)$ raised to the power $c\theta$, where $c$ known as the
"specific growth-rate constant" of the function, and, as mentioned
previously, $\theta$ is the angle of rotation of the radius $r_\theta$ (see figure 8.1A).

**210**

The consequences of exponential growth are often made more apparent by considering the derivative of the function. For equation 8.1, the derivative form is

$$\frac{dr}{d\theta} = cr \qquad (8.2)$$

Here one can see that the rate of change of the magnitude of the radius (i.e., the change in $r$ per change in rotation angle $\theta$) is a function of the radius itself, and the specific growth-rate constant of the function. The fact that the rate of change of $r$ is a function of the magnitude of $r$ itself leads to the characteristic increases in magnitude change that we know as "exponential."

FIGURE 8.1.  A: logarithmic spiral in two dimensional polar coordinates, in which the trace of the spiral is characterized in terms of the coiling angle $\theta$, the tangent angle $\alpha$, and radii $r$. B: the same spiral in two dimensional Cartesian coordinates, in which the trace of the spiral is characterized in terms of $x$-$y$ positional coordinates. C is a helicospiral in three dimensions in which the trace of the spiral is now characterized in terms of $x$-$y$-$z$ positional coordinates. D is a helicospiral that is not a continuous function but is composed of incremental vector components. *From McGhee 1991.*

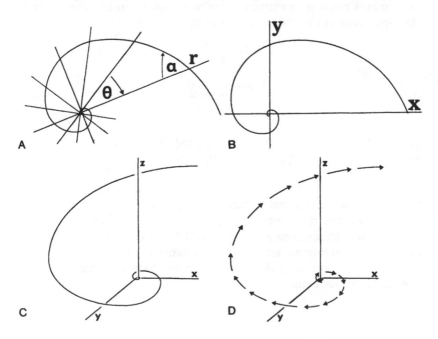

The rate at which the spiral "expands" is a function of the specific growth-rate constant $c$. For an actual logarithmic spiral, $c$ is usually measured by either the magnitude of the "tangent angle" to the spiral ($\alpha$) or the "whorl expansion rate" $W$ of Raup (1966), which he defined as the "ratio of two radii after one full revolution" (we encountered the "whorl expansion rate" many times in chapters 4 and 5, for both molluscs and brachiopods). The mathematical relationship of these two geometric parameters and the specific growth-rate constant of the exponential growth equation 8.1 is as follows:

$$c = \cot \alpha = \frac{(\ln W)}{2\pi} \tag{8.3}$$

From the previous equations, the logarithmic spiral can be characterized in terms of three geometric parameters: the rotation angle $\theta$, the tangent angle $\alpha$ (or whorl expansion rate $W$), and radii $r$ (figure 8.1A). These describe the logarithmic spiral in "polar coordinates," in which locations in polar space are described in radial magnitudes and polar angles. Most people, however, are more accustomed to thinking in terms of Cartesian coordinates (or at least I am). The logarithmic spiral can also be represented in Cartesian coordinates (see figure 8.1B). To produce this transformation, the previous single equation (8.2) in polar coordinates must be rewritten in terms of two simultaneous differential equations in Cartesian coordinates:

$$\frac{dx}{d\theta} = cx - y \tag{8.4}$$

$$\frac{dy}{d\theta} = cy + x \tag{8.5}$$

## NEXT: MOVING THE SIMULATION OF SHELL FORM INTO THE THIRD DIMENSION

Most of the spirals that we see in the coiled shells of molluscs and brachiopods are not simple logarithmic spirals but, rather, *helicospirals*, as discussed in chapter 4. This simply means that these spirals coil in three dimensions (see figure 8.1C). We can easily create helicospirals by adding an additional exponential into the third dimension. The growth rate of this new equation I have designated as $H$, or the "helicospiral component" (McGhee 1978a):

$$\frac{dz}{d\theta} = Hz \tag{8.6}$$

# CHANGING MODELING GOALS: THE SIMULATION OF THE PROCESS OF GROWTH ITSELF

The three equations 8.4, 8.5, and 8.6 are all that we need to model much of the spiral morphology seen in such diverse groups as molluscs, brachiopods, foraminifera, and so on. These equations all are continuous differentials, however. Growth in most organisms is not actually continuous but occurs in increments, even in organisms that have accretionary growth systems. If we wish to begin to model the growth process itself that produces the spiral morphology, it is biologically more meaningful to think in terms of a series of vectors in a spiral (figure 8.1D), where each vector represents a growth increment. Discrete growth can easily be modeled by rewriting the previous continuous differentials in terms of difference equations (Bayer 1977, Mc-Ghee 1978a):

$$\frac{(x_{j+1} - x_j)}{(\theta_{j+1} - \theta_j)} = cx_j - y_j \tag{8.7}$$

$$\frac{(y_{j+1} - y_j)}{(\theta_{j+1} - \theta_j)} = cy_j + x_j \tag{8.8}$$

$$\frac{(z_{j+1} - z_j)}{(\theta_{j+1} - \theta_j)} = Hz_j \tag{8.9}$$

This allows us to specify a third parameter, which I designate as $I$ for the growth "increment magnitude" (McGhee 1978a). This allows us to eliminate the parameter $\theta$, as only the difference magnitude is important:

$$I = \theta_{j+1} - \theta_j \tag{8.10}$$

The previous difference equations 8.7, 8.8, and 8.9 can now be simplified:

$$x_{j+1} = x_j + I(cx_j - y_j) \tag{8.11}$$

$$y_{j+1} = y_j + I(cy_j + x_j) \tag{8.12}$$

$$z_{j+1} = z_j + I(Hz_j) \tag{8.13}$$

The x-y-z coordinate of each new growth step, say step $j + 1$, is represented on the left side of the equations and is a function of the previ-

**213**

ous growth step, say step $j$, represented on the right side of the equations. We can further simplify these equations by factoring out all the growth parameters:

$$x_{j+1} = (1 + Ic)x_j - (I)y_j \tag{8.14}$$

$$y_{j+1} = (I)x_j + (1 + Ic)y_j \tag{8.15}$$

$$z_{j+1} = (1 + IH)z_j \tag{8.16}$$

and by rewriting these linear algebraic equations in terms of vectors and a matrix (see also Bayer 1978b):

$$\begin{bmatrix} x_{j+1} \\ y_{j+1} \\ z_{j+1} \end{bmatrix} = \begin{bmatrix} (1 + Ic) & (-I) & 0 \\ (I) & (1 + Ic) & 0 \\ 0 & 0 & (1 + IH) \end{bmatrix} \begin{bmatrix} x_j \\ y_j \\ z_j \end{bmatrix} \tag{8.17}$$

Growth increments $(j + 1)$ and $(j)$ are now given as two column vectors, and all the growth parameters are located in a single matrix. We may now rewrite equation (8.17) as

$$\vec{a}_{j+1} = \Lambda \vec{a}_j \tag{8.18}$$

Now the three linear algebra equations 8.14, 8.15, and 8.16 are simplified to a single matrix algebra equation (8.18). Each growth increment along the spiral is symbolized by a vector (figure 8.1D), which is a function of the previous growth increment (a vector) and the "growth-parameter matrix," or $\Lambda$.

The reader has probably noticed that in writing equation 8.18, I changed the notation that I have been using thus far in the book. Previously I followed standard mathematical practice by written scalars in italics (like $x$ or $r$) and vectors in boldface (like $\mathbf{v_0}$ or $\mathbf{v_T}$). In this chapter, I shall depart a bit from this practice by using an older notation in which a symbol with an arrow over it denotes a vector, as in equation 8.18. I do this here because we shall soon encounter vectors *that are themselves composed of vectors* (and not scalars), and I find the alternative notation serves as a useful visual reminder in those cases.

# NEXT: MAKING THE MODEL SIMULATION
# OF GROWTH EVEN MORE REALISTIC

A mollusc or brachiopod shell consists of numerous tiny growth lines. Each growth line, or growth increment, is composed of many different vectors, not just one (figure 8.2). To simulate a given growth increment, we specify two indices for the growth vector: $j$ for each growth step (as before) and $i$ for each separate growth vector (figure 8.2).

We can now go back and rewrite the original linear algebraic expressions (8.14, 8.15, and 8.16) to simulate many different spirals simultaneously:

$$x_{ij+1} = x_{ij} + I_i(c_i x_{ij} - y_{ij}) \qquad (8.19)$$

$$y_{ij+1} = y_{ij} + I_i(c_i y_{ij} + x_{ij}) \qquad (8.20)$$

$$z_{ij+1} = z_{ij} + I_i(H_i z_{ij}) \qquad (8.21)$$

Following our previous procedure (rewriting equations 8.14, 8.15, and 8.16 in terms of equations 8.17 and 8.18), these linear expressions can now be written in matrix form:

$$
\begin{bmatrix} \vec{a}_{1j+1} \\ \vec{a}_{2j+1} \\ \vec{a}_{3j+1} \\ \vdots \\ \vec{a}_{nj+1} \end{bmatrix}
=
\begin{bmatrix}
\Lambda_1 & 0 & 0 & \cdots & 0 \\
0 & \Lambda_2 & 0 & \cdots & 0 \\
0 & 0 & \Lambda_3 & \cdots & 0 \\
\vdots & \vdots & \vdots & & \vdots \\
0 & 0 & 0 & \cdots & \Lambda_n
\end{bmatrix}
\begin{bmatrix} \vec{a}_{1j} \\ \vec{a}_{2j} \\ \vec{a}_{3j} \\ \vdots \\ \vec{a}_{nj} \end{bmatrix}
\qquad (8.22)
$$

Each vector is now, however, a vector composed of many smaller vectors, and the central matrix is itself a matrix of smaller matrices located along the principal diagonal. In matrix terms, the complete simulation of a growth line or increment in a shell may be written as

$$\mathbf{a}_{ij+1} = \Lambda_i \mathbf{a}_{ij} \qquad (8.23)$$

where $\mathbf{a}_{ij}$ is a vector composed of vectors, and $\Lambda_i$ is a matrix composed of matrices. Each growth line or increment $\mathbf{a}_{ij+1}$ is a function of the previous growth line $(\mathbf{a}_{ij})$ and the developmental matrix $(\Lambda_i)$.

FIGURE 8.2. A hypothetical mollusc shell, with a single growth increment along the margin of the shell viewed as composed of many separate growth vectors. *From McGhee 1991.*

# MODELING ANISOMETRY: ONTOGENETIC ALTERATIONS IN GROWTH PARAMETERS

Let us return for a moment to simple exponential growth systems, as typified by the logarithmic spiral (equations 8.1 and 8.2). One of the fundamental properties of the logarithmic spiral is that it remains the same shape, regardless of its size, and thus, as stated earlier in the chapter, actually simulates an isometric growing system. However, most growth in nature is not isometric. Raup's (1966, 1967) early work with shell form in molluscs has been criticized somewhat unfairly for the assumption of isometry (see the discussion in chapter 4), but it is a fact that anisometric growth is common even in molluscs.

Earlier we examined $c$, the "specific growth-rate constant" of an exponential function (equations 8.1 and 8.2). Because $c$ is a constant, the tangent angle of the logarithmic spiral $(\alpha)$ and the whorl expansion rate of the logarithmic spiral $(W)$ also are constants (see equation 8.3). If we use the logarithmic spiral to simulate growth in snails, for example, the whorl expansion rate of the snail's shell never changes during its life, and the shape of its shell also never changes, regardless of its ultimate size. What if this were not true, and the shape of the snail's shell did change markedly during its life, that is, if the snail experienced anisometric growth? Then obviously the whorl expansion rate of its shell would also have to change during growth and would no longer remain a constant.

Ontogenetic changes in shape require that the specific growth rate be treated as a variable rather than a constant. Change in the specific growth rate can be expressed as function of the age of the animal under consideration or, in the logarithmic spiral model, as a function of the polar angle:

$$c = 2\pi^{-1}(\ln W) = f(\theta) \tag{8.24}$$

Notice that in equation 8.24, $c$ is now no longer a constant, as in equations 8.2 and 8.3, but is a *function*. Applying Taylor's theorem, $f(\theta)$ may be expanded as a power function in $\theta$.

$$\ln W = 2\pi c_1 + 2\pi c_2\theta + 2\pi c_3\theta^2 + 2\pi c_4\theta^3 + \cdots \tag{8.25}$$

The simplest assumption to make at this point is that the specific growth rate is modified as a constant rate during ontogeny—that is, that the logarithm of the whorl expansion rate is a linear function of

**217**

$\theta$ (remember in modeling always to start with the simplest assumptions first):

$$\ln W = 2\pi c_1 + 2\pi c_2 \theta \qquad (8.26)$$

The whorl expansion rate of our hypothetical snail is now a function of its age (as measured by $\theta$ in the model). If $W$ changes with time, we can introduce a new parameter to measure the rate of that change. This parameter I have termed $M$, for the "specific whorl expansion modification rate" (McGhee 1980b). It is defined as the derivative of equation 8.26:

$$M = \frac{dW}{Wd\theta} = 2\pi c_2 \qquad (8.27)$$

The logarithmic spiral model is given in equation 8.1. What, then, is form of this new model, one in which the whorl expansion rate is a function of $\theta$ rather than a constant? To find out, we can rewrite equation 8.26 solely in terms of $r$ and $\theta$ by using the relationships given in equation 8.2 and 8.3:

$$\frac{dr}{d\theta} = r(c_1 + c_2\theta) \qquad (8.28)$$

where again one can see that the rate of change of $r$ is now no longer simply $r$ multiplied by a constant (as in equation 8.2), but $r$ multiplied by a function of $\theta$. We can then integrate equation 8.28:

$$\int r^{-1}dr = \int (c_1 + c_2\theta)d\theta \qquad (8.29)$$

which, setting $\ln r_0$ equal to the constant of integration, yields

$$\ln r_\theta = \ln r_0 + c_1\theta + \left(\frac{1}{2}c_2\right)\theta^2 \qquad (8.30)$$

which, removing the logarithms, simplifies to

$$r_\theta = r_0 e^{c_1\theta + \left(\frac{1}{2}c_2\right)\theta^2} \qquad (8.31)$$

and gives us the form of the new model.

At this point I can hear the reader asking, does any of this modeling have any practical relevance to reality? Certainly! Let us consider the case of *Lycophoria nucella*, a very real brachiopod species, whose individuals show marked anisometric growth. First, however, let us quickly reexamine equation 8.2, the derivative of the logarithmic spiral model (equation 8.1) in which no shape change occurs with ontogeny. If we integrate equation 8.2, as we did with equation 8.28 (resulting in equation 8.30), we get

$$\ln r_\theta = \ln r_0 + c\theta \qquad (8.32)$$

Thus, if we measure successive radii and angles from the spirals seen in the dorsal and ventral valves of the brachiopod *Lycophoria nucella* and if we plot the logarithms of $r$ versus $\theta$ in a graph and that plot turns out to be linear, it means that the spirals seen in *Lycophoria nucella* valves are logarithmic and that individuals of this long extinct species grew isometrically. If the plot turns out not to be linear, however, the brachiopods grew anisometrically.

Figure 8.3 gives an actual plot of $r$ and $\theta$ measurements taken from the dorsal valve of an individual of the species *Lycophoria nucella*. Two functions were fit to those data by means of regression, first a linear function (equation 8.32, which indicates isometric growth) and second a quadratic function (equation 8.30, which indicates anisometric growth). The quadratic function clearly (and significantly, in the statistical sense) fits the measurement data better than the linear function does; thus this individual of *Lycophoria nucella* experienced anisometric growth in its lifetime.

But there is more than just anisometry. If the whorl expansion rate changes during growth, rather than remaining constant, the value of $W$ may either increase or decrease with the age of the animal. In the model, the sign of $M$, the whorl expansion modification rate (equation 8.27), may be either positive or negative. If positive, the magnitude of $W$ will increase during growth; if negative, it will decrease (and if $M$ is zero, growth will be isometric). For *Lycophoria nucella*, the sign of $M$ is negative (figure 8.3), indicating that $W$ decreased during growth and that the convexity of the shell increased with age.

Study has shown that *Lycophoria nucella* is not an isolated case of anisometric growth involving decreasing $W$ values during ontogeny in biconvex brachiopods. The great majority of brachiopods that exhibit anisometric growth show decreases in $W$ with time; that is, the shell of the animal becomes more and more convex with age (McGhee

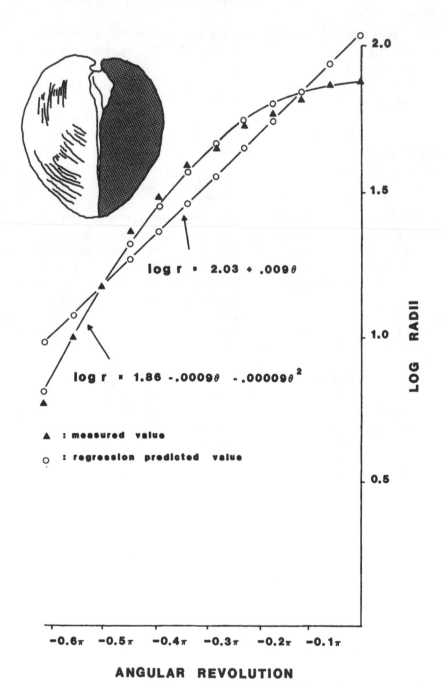

log r = 2.03 + .009θ

log r = 1.86 - .0009θ - .00009θ²

▲ : measured value

○ : regression predicted value

1980b). Figure 8.4 gives the ontogenetic trackways, or growth trajectories, of five species of brachiopods in the dorsal-$W$–ventral-$W$ theoretical morphospace discussed in chapter 5 (see figure 5.1). All the brachiopods move from the upper right to the lower left in the morphospace during ontogeny, regardless of whether the ontogenetic trackway is or long or short (i.e., whether the reductions in $W$ are large or small during ontogeny).

Why do biconvex brachiopods decrease $W$ values in the valves of the shell with growth? In chapter 5 I argued that biconvex brachiopods tend to produce shells that have maximum internal volumes to minimum shell surface areas; that is, they approach a spherical form as closely as possible within specific geometric constraints. Brachiopods are prevented from reducing the magnitude of $W$ in the valves to values below $10^2$ by limitations imposed by their system of articulating the two planispirally coiled valves into a single functional bivalved shell (see the discussion in chapter 5). Yet figure 8.4 shows four brachiopod species whose ontogenetic trajectories have clearly entered the forbidden zone of $W$ values of less than $10^2$ (cf. figures 5.11 and 8.4). Many brachiopods have found a way around the articulation limitations imposed by the necessity of avoiding whorl overlap in both valves, by growing anisometrically. They begin life with shells with high $W$ values, far from the region of whorl overlap shown in figure 5.1, and then progressively decrease the magnitude of $W$ in the valves with growth. The brachiopod shell thus starts out flattish in the critical posterior region of the shell where the articulation must be maintained but then becomes more convex with time and eventually approaches a spherical shape in the later anterior growth stages, as can be seen in the profile sketch of *Lycophoria nucella* in figure 8.3.

The analytic technique outlined here (and in McGhee 1980b) was also used by McKinney (1984) in his analysis of allometry in echinoids and by Ackerly (1992b) in his study of ontogenetic variation in shell morphology in the bivalve mollusc *Pecten*, and a slightly modified form of equation 8.30 was used by Schindel (1990) to study anisometric growth in gastropods. Thus, to answer the hypothetical reader's

FIGURE 8.3. Successive linear and quadratic functions fit to measurements taken from the dorsal valve of the pentameride *Lycophoria nucella*. Measurement data are indicated by solid triangles, and regression predicted values, by open circles. The addition of the second-order polynomial produces a significant reduction in variance about the regression function. Values of θ are given in radians. *From McGhee 1980b.*

earlier question, theoretical morphologic modeling clearly has practical values as well as being fun!

Equations 8.24 to 8.31 are modifications of the logarithmic spiral model to produce a model that allows for anisometric growth. Alternatively, we can directly model anisometric growth itself, rather than modifying a previously existent isometric model to allow for anisome-

FIGURE 8.4. Examples of ontogenetic alterations of whorl expansion rates for five species of pentameride brachiopods. Arrows indicate the measured direction of the ontogenetic trajectories in the theoretical morphospace. In all cases, the trajectories are from higher to lower values of $W$ in both valves. Species are indicated by numbers: (1) *Virgiana barrandei*, (2) *Lycophoria nucella*, (3) *Conchidium biloculare*, (4) *Pentameroides subrectus*, and (5) *Stricklandia (Stricklandia) lens*. Numbers in circles indicate the average whorl expansion rates for the ontogeny of the species individuals. *From McGhee 1980b.*

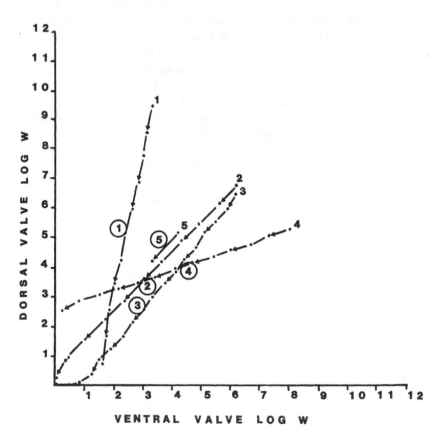

try. Let us return to equation 8.23, in which each growth increment in a shell is defined as a function of the previous growth line and the developmental matrix $\Lambda_j$. Equation 8.23 simulates a vector field for all positions along the growing edge $(i)$ at any one point in time $(j)$. The developmental matrix, of course, could *change* from each $j$th growth increment to the next, thus allowing us to simulate ontogenetic changes in the growth parameters comprising the matrix. If we go back to the equation for a single growth spiral (equation 8.18), a formal representation of ontogenetic alteration of $\Lambda$ for each $j$th increment along the spiral can be written (see Bayer 1978b):

$$
\begin{aligned}
\vec{a}_1 &= \Lambda_1 \vec{a}_0 \\
\vec{a}_2 &= \Lambda_2 \vec{a}_1 = \Lambda_2 (\Lambda_1 \vec{a}_0) \\
\vec{a}_3 &= \Lambda_3 \vec{a}_2 = \Lambda_3 (\Lambda_2)(\Lambda_1 \vec{a}_0) \\
\vdots &= \vdots
\end{aligned}
\tag{8.33}
$$

which simplifies to the following expression:

$$
\vec{a}_n = \prod_{j+1}^{n} \Lambda_j \, (\vec{a}_0)
\tag{8.34}
$$

Equation 8.34 now can be used to simulate ontogenetic changes in the vector field, as determined by changes in the developmental matrix $\Lambda_j$ with time.

## SUMMARY: A DISCUSSION OF THE THEORETICAL MORPHOLOGY EXAMPLE

Several features of the process of theoretical morphologic modeling can be seen in the example given in this chapter. The first was to examine the geometric properties of the logarithmic spiral (figure 8.1A and B) in equations 8.1 to 8.5, where we see that such a spiral is produced by an exponential growth process and, moreover, one in which growth is isometric. Isometric growth, in which the proportion does not change with increasing size, follows from the fact that the growth rate of the function (whether measured by $c$, $W$, or $\alpha$) is a constant and thus does not vary with time (or size). Therefore, the logarithmic spiral can be used only to simulate form in those organisms that grow isometrically.

Second, the logarithmic spiral exists only in two-dimensional space (figure 8.1A and B). Even in those organisms that coil planispirally (brachiopods, most ammonites, some gastropods, and so on), there is actually only one section through the shell or valve of the organism that contains a true logarithmic spiral in two dimensions (the "directive spiral" of Stasek 1963). All other spirals in the shell or valve coil in three-dimensional space and are helicospirals (figure 8.1C), and we can easily model these in equation 8.6.

Simultaneously with our modifying the original model to reflect reality more closely, we have begun parameterizing form. By equation 8.6, we are prepared to describe organisms with coiled shells in terms of two parameters: the growth-rate constant $c$ and the helicospiral component of growth $H$. More complex characterizations of form require more parameters. One of the goals of theoretical morphology is, however, to minimize the number of parameters needed to begin exploring the range of form potentially available in nature. Raup (1966) created the first hypothetical morphospace for univalved shell geometries using only three parameters, and his approach revolutionized the way we view the distribution of form in ammonites and gastropods.

The purpose of the theoretical exercise shifts in equations 8.7 through 8.9. Previously, we simulated form using a geometric model. In equations 8.7 through 8.9, we began to simulate the process of growth itself. Examination of the shells of molluscs reveals that they are composed of numerous increments of shell accretion (figure 8.2), as reflected in the growth lines. If we wish to simulate the process of accretionary growth that produced the shell, we must abandon continuous differential equations and switch to difference equations. This allows us to specify a new parameter, $I$, to simulate the size of the growth increments in the model (equations 8.10 through 8.13) and to redefine the model in terms of matrix algebra (equations 8.14 through 8.18).

Growth along the margin of a mollusc shell is produced by the activity of numerous cells in the mantle margin. In most simulations of three-dimensional shell growth, the growth margin is represented by a geometric figure, usually a circle or ellipse (the "generating curve" in Raup 1966), which grows (expands) according to the mathematics of the model. To simulate the morphogenetic process of cellular accretion along a mantle margin, geometric generating curves must be abandoned and individual growth vectors along the margin substituted instead, as in equations 8.19 through 8.21. In the final expression, the resulting model contains vectors *that are themselves composed of vectors* and a matrix *that consists of other matrices* (equations 8.22 and 8.23).

We can likewise simulate anisometric growth with a geometric model or simulate the anisometric growth process itself. If we wish to model anisometric growth with the logarithmic spiral model, we must modify the model so that the growth rate is no longer a constant but, rather, a function of growth (equations 8.24 through 8.31). Such a model has been demonstrated to be useful in the actual analysis of anisometric growth in brachiopods (McGhee 1980b), echinoids (McKinney 1984), gastropods (Schindel 1990), and bivalve molluscs (Ackerly 1992b). Alternatively, we can use the difference model of actual incremental growth (equation 8.18) to simulate ontogenetic changes in the growth parameters (equation 8.33 and 8.34).

We started the chapter with a model of form, equation 8.1, and we finish with a model of morphogenesis, equation 8.34. Both types of models are understood today as theoretical morphology. And all are equally fun to experiment with.

# IX

# Theoretical Models of Accretionary Growth Systems

*The use of models such as that developed here is a more powerful tool than statistical techniques such as multivariate analysis. Statistical procedures are useful for obtaining empirical generalizations . . . nevertheless, empirical generalizations have to be explained by means of higher level hypotheses or theories. A mathematical model is a higher level hypothesis.*
—*De Renzi (1988:399)*

Most theoretical models of accretionary morphogenesis to date have been designed to analyze invertebrate shell form in nature. This has been done largely because many invertebrate organisms that produce shells grow isometrically and thus can be modeled with simple exponential growth equations, such as the logarithmic spiral, as we saw in the previous chapter. Spiral geometries are exceedingly abundant in nature, however, and can be found in many aspects of organic morphology other than shells, not just in animals but in plants as well and also in organisms that grow anisometrically (Stevens 1974).

In general, theoretical morphogenetic models may simulate either organic form or the growth process itself that produces the organic form under study. The former are usually written in terms of continuous differential equations, and the latter are usually written in terms of incremental difference equations.

There are two general classes of theoretical morphologic models of shell form: those that use a generating curve and those that do not. Models that use a generating curve have been variously described as "growing-tube models" (Okamoto 1988a:39), "tube models" (Checa 1991:97), and "helicoconical-tube models" (McGhee 1991:93). They will be termed "generating-curve models" here, as the essential feature is the initial presence of a generating curve that is then allowed to "grow" and produce a helicoconical-tube shell simulation. In contrast, models that do not use a generating curve will be termed "multivector models" (Bayer 1978b, McGhee 1978a).

Second, "growth" in theoretical morphogenetic models may take place either relative to a fixed-reference frame or not. Models that do not have a fixed-reference frame are described as "moving-reference frame" models (Ackerly 1989a; Okamoto 1988a).

## INCREMENTAL GROWTH MODELS
## WITH DIFFERENCE EQUATIONS

In the mathematical example given in chapter 8, it was pointed out that accretionary growth is an incremental process, yet the shell forms produced by this process are generally modeled by continuous differential equations. If we wish to model morphogenesis, it is biologically more meaningful to switch from continuous differentials to difference equations, as we did in equations 8.7 to 8.9.

In some cases, the morphology produced by difference equations can differ markedly from that produced by differentials with the same parameters, particularly if the increment magnitude is large (see equations 8.10 through 8.13). The biological significance of modeling morphogenesis by difference equations was discussed in some detail by Bayer (1977, 1978b) and McGhee (1978a). Difference equations with incremental growth steps can easily be incorporated into theoretical morphologic models as diverse as the multivector type (Bayer 1978b; McGhee 1978a) or generating-curve models with moving-reference frames (Ackerly 1989a, Savazzi 1990a), as we shall see in this chapter.

## GENERATING-CURVE MODELS WITH
## FIXED-REFERENCE FRAMES

*The surface of any turbinated or discoid shell may be imagined to be generated by the revolution about a fixed axis (the axis of the shell) of the perimeter of a geometrical figure, which, remaining always geometrically similar to itself, increases continually its dimensions.*

*Moseley (1838:351)*

A point is a point. If a point moves through space with time, its trajectory will describe a line. If that line has a curvature that never changes, a logarithmic spiral will result (as in figure 8.1 or 8.2). A gastropod shell is a three-dimensional object, however, not a line. The key to the theoretical simulation of shell form was proposed over a century and a half ago by Moseley (1838): rather than moving a point through space, move a "geometrical figure" instead. Moseley variously termed this geometric figure a "generating figure" and a "generating curve," as in "the generating figure of the *Conus virgo* is a triangle, that of the *Trochus telescopicus* and of the *Trochus archimedis*, a trapezoid" (Moseley 1838:351). In his review of Moseley's work, D'Arcy Thompson chose to use the term "generating curve" rather than "generating figure," as in "let us imagine a closed curve in space, whether circular or elliptical or some other and more complex specific form, not necessarily in a plane: such a curve as we see before us when we consider the mouth, or terminal orifice, of our tubular shell. Let us call this closed curve the 'generating curve'" (Thompson 1942:778–779). The term "generating curve" was adopted by Raup in his initial paper on theoretical morphologic modeling, though he did mention that the generating curve "may be thought of, in most instances, as the whorl cross section" (Raup 1961:603). Thus the term "generating curve" is firmly entrenched in the literature of theoretical morphology, even though the term is somewhat confusing, as most people think of a "curve" as a segment of a geometric figure and not an entire figure such as a circle or an ellipse. (A brief aside: instead of using a circle or an ellipse as a generating curve, Pickover's simulation technique [1989, 1991] uses a *sphere* as a generating *figure*. The centers of the spheres all fall along a helicospiral trajectory in space, where the spheres expand in size and interpenetrate one another with time, producing some very interesting computer-simulated morphologies. The simulated morphologies are biologically unrealistic as seashells, however, in that they have no apertures. They actually appear more similar to certain plant fruit structures, such as some gourds or squash.)

All three essential elements of generating-curve models with fixed-reference frames are found in the single sentence by Moseley (1838) quoted in the beginning of this section of the chapter. These elements are (1) the generating curve, (2) the movement of the generating curve with respect to a fixed-reference frame, and (3) the specification of a mathematical function describing the trajectory of the generating curve through space. In Moseley's example, the mathematical function is seen in his specification that the hypothetical shell be produced by revolving the generating curve around a coiling axis so that the shell "increases continually its dimensions" while at the same time "remaining always geometrically similar to itself"; that is, the function is a logarithmic spiral.

Simulations of shell form using generating-curve models with fixed-reference frames have a long classical history (Moseley 1838, 1842; Thompson 1914, 1942; Lison 1942, 1949; Fukutomi 1953; Stasek 1963). These early shell form simulations are simple line drawings, produced by hand. Even so, many are very elegant and rival those produced today by computer; see, for example, the planispiral to helicospiral shell form simulations given in figure 29 of Lison (1949) and in figure 2 of Stasek (1963). Moseley (1838) also developed equations for calculating surface areas and volumes of logarithmically coiled shells, equations that inspired theoretical morphologists more than a century later (Raup and Chamberlain 1967, Raup and Graus 1972, Davaud and Wernli 1974, Stone 1997a).

Raup (1961, 1962, 1966, 1967) introduced two new elements to generating-curve models with fixed-reference frames: (1) simulation computations and graphics produced by computer and (2) formal parameterization of the mathematical function describing the trajectory of the generating curve through space. Even here, the parameters introduced by Raup ($W$, $D$, $T$, and $S$; see chapter 4 for discussions of each) are not new. The concept of a "translation rate" parameter can be found in Moseley (1838) and the concept of a variable "shape of the generating curve" parameter in Moseley (1838) and Thompson (1917, 1942). The effect of a "distance of the generating curve from the coiling axis" parameter can be seen in the simulations given in Lison's figure 28 (1942). Raup's parameters are also a geometric and morphogenetic mixture; that is, some simply describe form, and others simulate the process that generates that form. Subsequent workers have reformulated the logarithmic spiral model in terms of parameters that are more "biological" and refer directly to the growth process itself (Løvtrup and Løvtrup 1988; Løvtrup and von Sydow 1974, 1976; Rice 1998). What is different about Raup's parameters, other than their formalization, is how they were used.

To illustrate this difference, let us consider the geometric parameter $W$, the whorl expansion rate. Raup (1961:604) initially defined this parameter as the ratio of whorl widths "where successive measurements are separated by a full revolution about the axis" (a more general expression of $W$ for angles less than $2\pi$ is given in equation 5.1 of this book). This same geometric parameter was proposed more than 120 years earlier by Naumann (1840a), who called it the *Windungsabstände* ("whorl spacings") and designated it as $q$, since he usually expressed it as a quotient. If we compare the equation given at the top of page 232 in Naumann (1840a) for calculating a radius in a logarithmic spiral,

$$r = aq^{v/2\pi} \tag{9.1}$$

with that given in Raup (1966:equation 1),

$$r_\theta = r_0 W^{\theta/2\pi} \tag{9.2}$$

we will see that they are in fact the same. The polar angle is designated as $v$ in Naumann (1840a) and as $\theta$ in Raup (1966), and the initial radius (at $v$ equal to zero) is $a$ in Naumann (1840a) and $r_0$ in Raup (1966); otherwise the two expressions are identical.

All similarity ends, however, when the use of this parameter is contrasted between Raup and Naumann. Raup (1961, 1962) used it as a theoretical parameter to *simulate hypothetical shell forms* and finally to create the *W–D–T* theoretical morphospace discussed in chapter 4 (Raup 1966, Raup and Michelson 1965). Naumann (1840a,b) used it as a morphometric parameter to *describe existent shell form* and to test how closely actual morphologies approximate a logarithmic spiral, by comparing observed measurements with predicted values for a series of gastropods and ammonites. Naumann (1840a,b, 1845) used this parameter to initiate a debate with Moseley (1838, 1842) over the actual utility of the logarithmic spiral in describing real morphology, with Naumann (and his students) proposing more complicated models in quest of *die wahre Spirale der Ammoniten* ("the true spiral of the ammonites"). But such a quest is morphometrics, not theoretical morphology (as discussed in chapter 1) and so will not be pursued any further here. The *Windungsabstände* parameter had to wait 120 years to move from the field of morphometrics to that of theoretical morphology.

The mathematical function most frequently used to specify the trajectory of the generating curve though space in fixed-reference frame models is the logarithmic spiral, discussed in chapter 8 (see equations 8.1 through 8.6). The logarithmic spiral exhibits the interesting geometric property of gnomonic growth, that is, that changes in size occur without entailing any change in proportion. In accretionary growth systems, each growth increment added is a gnomon to the rest of the shell, and so growth does not alter the shape of the generating curve. Such a growth system is isometric, and although isometric growth is found in many organisms using accretionary growth systems, it is the exception rather than the rule. Most growth systems in nature are anisometric.

The logarithmic spiral model can be modified to simulate anisometric growth (McGhee 1980b, Bayer 1985, Cortie 1989), as shown in chapter 8 (see equations 8.24 through 8.31). Anisometric modifications of the logarithmic spiral model were demonstrated to simulate the anisometric growth patterns found in brachiopods (McGhee 1980b), bivalves (Ackerly 1992b), and gastropods (Schindel 1990). Computer simulation of a shell form showing anisometric growth, produced by modifications of the logarithmic spiral model, is given in text-figure 5 in Raup's early work (1966).

A particularly nice set of simulations showing an anisometric growth series of shell forms, again produced by modifying the logarithmic spiral model, were generated by Cortie (1989) and are included here in figure 9.1. Cortie (1989, 1993) produced a much more intricate model, one containing 16 parameters:

$$x = D[A\sin \beta\cos \theta + R\cos (s + \phi) \cos (\theta + \Omega) \\ - R \sin \mu\sin (s + \phi)\sin \theta]e^{\theta\cot \alpha} \tag{9.3}$$

$$y = [-A\sin \beta\sin \theta - R\cos (s + \phi) \sin (\theta + \Omega) \\ - R \sin \mu\sin (s + \phi) \cos \theta]e^{\theta\cot \alpha} \tag{9.4}$$

$$z = [-A\cos \beta + R \sin(s + \phi) \cos \mu]e^{\theta\cot \alpha} \tag{9.5}$$

where $D$ specifies the direction of coiling (dextral or sinistral); $A$ and $R$ specify various spatial locations for the aperture; $\beta$, $\phi$, $\Omega$, $\mu$, and $s$ specify various tilts and rotations of the aperture; and $\alpha$ and $\theta$ are the familiar tangent angle of the spiral and polar angle of the spiral, respectively. Although much more complicated than the simple logarithmic spiral model, the end term $(\exp \theta \cot \alpha)$ in each of the three equations (9.3, 9.4, and 9.5) is the same as that in equations 8.1 and 8.3 (a brief aside: equations 9.3 and 9.4 were misprinted in Cortie 1989, missing the first $\theta$ term. The equations are correct in Cortie 1993).

Cortie's (1989, 1993) model is capable of simulating shell surface ornamentation (spines, knobs, and so on) as well as various orientations of the shell aperture not possible with Raup's simple four-parameter model (1966). Variations in apertural orientations are particularly important to the analysis of actual shell form in gastropods, and so their absence in Raup's $W$–$D$–$T$ theoretical morphospace was an impediment to using that morphospace to explore the range of helicospiral shell form found in nature (see chapter 4).

Another complexity in generating-curve models with fixed-reference frames can be found in the work of Fowler, Meinhardt, and Prusinkiewicz (1992) and Prusinkiewicz and Fowler (1995), who added

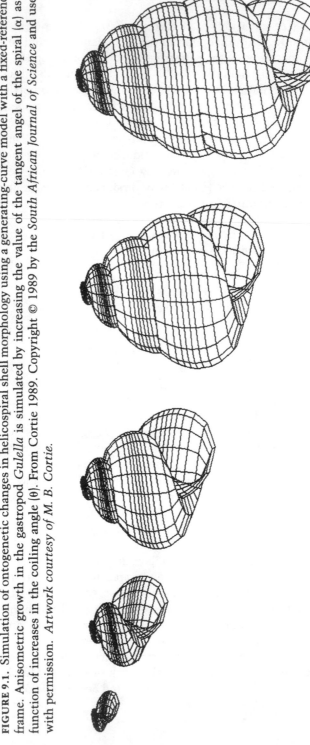

**FIGURE 9.1.** Simulation of ontogenetic changes in helicospiral shell morphology using a generating-curve model with a fixed-reference frame. Anisometric growth in the gastropod *Gulella* is simulated by increasing the value of the tangent angel of the spiral (α) as a function of increases in the coiling angle (θ). From Cortie 1989. Copyright © 1989 by the *South African Journal of Science* and used with permission. *Artwork courtesy of M. B. Cortie.*

simulation of the intricate external color patterns found on many molluscan shells to the model of shell form simulation itself. Accordingly, their model is a wedding of "generating-curve models with fixed-reference frames" to "reaction-diffusion models" (more on these in the next chapter).

Although the logarithmic spiral model has been the most frequently used to specify the trajectory of the generating curve through space, other mathematical functions can be used in its place. In a logarithmic spiral, increases in magnitude of the radius follow an exponential function (see equation 8.1). Rather than exponential growth, the radius might grow in a linear function:

$$r = a\theta + b \qquad (9.6)$$

where the rate of growth of the radius is a constant

$$\frac{dr}{d\theta} = a \qquad (9.7)$$

rather than a function of $r$ itself, as in the case of the logarithmic spiral (see equation 8.2). Such a linear growth system produces an *Archimedes spiral*, in which each of the whorls in the spiral morphology has the same width, rather than expanding in width, as in the logarithmic spiral.

Another possibility is that growth might proceed in an allometric fashion; that is, growth could follow a power function:

$$r = a\theta^n \qquad (9.8)$$

where the linear growth model of equation 9.6, and thus the Archimedes spiral, can be taken as a special case of the power function given in equation 9.8, that is, where the value of $n$ is equal to one. In a power function, the rate of growth of the radius is a function:

$$\frac{dr}{d\theta} = na\theta^{n-1} \qquad (9.9)$$

and not a constant, as in equation 9.7. The growth rate of the radius, however, is now not a function of $r$ itself, as in exponential growth (see equation 8.2) but, rather, of the polar angle $\theta$. The usage of allometric, rather than exponential, functions to model shell form (see

Burnaby 1966) is almost as old as the original work of Raup himself (who used exponential functions; see Raup 1961, 1962, 1966). More recently, Stone (1995) developed an interesting generating-curve model with fixed-reference frames that uses allometric functions to model magnitude changes in aperture trajectory and scaling in the computer simulations of shell form, and he also used the model to develop a "hybrid morphospace" (see chapter 2) to reexamine evolutionary trends in the modern gastropod genus *Cerion* (Stone 1996a).

Alternatively, the growth rate of the radius might still be a function itself, but a more complex function than the exponential (and the logarithmic spiral). Growth could follow a logistic function instead:

$$\frac{dr}{d\theta} = kr\left(1 - \frac{r}{r_m}\right) \qquad (9.10)$$

where $k$ is the growth-rate constant and $r_m$ is the limit of $r$. In such a growth system, the rate of increase in the radius accelerates to a maximum value and then decelerates as the magnitude of the radius approaches its limit. De Renzi (1988, 1995) used logistic growth functions to simulate the spiral morphologies found in the larger foraminifera like *Nummulites*. The logistic model is explicitly an anisometric model of growth and thus requires no further modification to simulate ontogenetic changes in form and proportion. Because the limit of $r$ is a function of the polar angle $\theta$, the rate of change of $r$ is a function of both $r$ and $\theta$, and so this anisometric growth model (equation 9.10) and that developed in chapter 8 (equation 8.28) are similar in this respect.

A totally different approach to specifying the mathematical function describing the trajectory of the generating curve through space was taken by Illert (1983, 1987, 1989), who proposed a "clock-spring" tensional model in which shell morphogenesis occurs "in accordance with Hooke's law of elasticity, in energy-efficient optimally strong geometries" (Illert 1987:793). Illert characterized a logarithmic spiral surface in terms to two vectors and a rotation matrix:

$$\mathbf{r}(\theta,\phi) = e^{\alpha\phi}\begin{pmatrix} \cos\phi & -\sin\phi & 0 \\ \sin\phi & \cos\phi & 0 \\ 0 & 0 & 1 \end{pmatrix} \mathbf{r}(\theta,0) \qquad (9.11)$$

similar to the linear mappings used by Bayer 1978b (compare Illert 1983:equation 1, given here as equation 9.11, with Bayer 1978b:equation 22). From this Illert derived a "symmetry equation" (Illert 1983:

equation 27), whose general solution is the same as we saw in chapter 8 (compare equations 39i, 39ii, and 39iii in Illert 1983 with equations 8.4, 8.5, and 8.6 in chapter 8 and equations 3, 4, and 5 in McGhee 1978a). At this point, Illert departed from conventional modeling and defined these differential equations in terms of an "energy functional," whose solution is argued to be the same as "Hooke's law for planar elastic spiral springs" (Illert 1983:45). In practice, the trace of the spiral function through space is considered to be analogous to an infinitely thin steel wire that is fixed at one end to a rigid tie point and may be wound at the other end like a clock spring (Illert 1987:figure 1). Later Illert (1989) added Frenet coordinates to the clock-spring wire simulation to create three-dimensional surfaces, and thus his later modeling falls into the "moving-reference frames" category of generating-curve models (where we shall again encounter Frenet frames). Proclaiming that he had provided "the solution of the classical seashell problem," Illert also posed interesting philosophical questions concerning "life forces" (Illert 1983:47) and commented on the historical development of "clockspring physics" in morphology (Illert 1987:792–793). In a later work, he (1991) advanced the idea that certain invertebrate animals may be either able to foretell the future or influence the past: "We are talking here, about *action with foreknowledge*, action outside the expected linear *Newtonian* time sequence, rather as if *an impending future event acted BACKWARD THROUGH (future) TIME to influence the present!*" (Illert 1991:455; italics and capitals Illert's).

Other morphologists are not as confident in the utility of clockspring physics, however, and Savazzi (1995:235) in particular stated that "no biological justification is available for accepting Illert's method as realistic. . . . In fact, clocksprings are static-equilibrium structures. A dynamically growing shell has no obvious analogue" (see, in contrast, Silk and Hubbard's [1991]model of twining in morning glories, which uses a "wire rope" and Frenet frame approach to study force balances in vines twining around poles). Savazzi (1995:235) further stated, however, that "in spite of the shortcomings of the clockspring method, it cannot be excluded that further work will reveal some useful properties of this method," and he did include the model in his published computer programs of theoretical morphology (Savazzi 1993).

At the beginning of this section, we listed the three essential elements of generating-curve models with fixed-reference frames: (1) the generating curve, (2) the movement of the generating curve with respect to a fixed-reference frame, and (3) the specification of a mathematical function describing the trajectory of the generating curve through space. In the next section, we shall examine models that have (1) and (3), but not (2); that is, they have no fixed frame of reference. For the reader who wishes to experiment further with generating-curve models with fixed-reference frames, see the "Class Raup" and the "Class Illert" models in the computer programs published by Savazzi (1993).

## GENERATING-CURVE MODELS WITH MOVING-REFERENCE FRAMES

Actual accretionary growth in organisms does not take place relative to a hypothetical reference frame, but rather, each increment of growth takes place relative to the previous growth increment. In many organisms, each accretionary addition to the shell is indeed isometric (a "gnomon"), so that the resultant shell can realistically be simulated by cones coiled logarithmically around a reference axis.

Fixed-frame models are quite limited in their ability to simulate organisms that grow anisometrically. Shells of such aberrant morphology as seen in vermetid gastropods and heteromorph ammonites may be simulated only with major (and often biologically unrealistic) modifications of the growth parameters if growth is to take place relative to a hypothetical coiling axis.

A significant improvement over the limitations inherent in fixed-frame models is provided by moving-frame models, in which the frame of reference for growth moves with the generating curve itself (Okamoto 1984; 1988a,b,c; 1993; Ackerly 1989a). In Okamoto's model (1988a), both the amount of growth of the generating curve and the trajectory of the generating curve through space are computed with reference to the previous position and size of the generating curve, and not to an external coordinate system. Okamoto's generating curve is always a circle with a radius $r$, and growth of the generating curve is measured in terms of increasing magnitudes of this radial vector. A second vector is oriented normal to the plane of the circular generating curve and intersects the plane of the circular generating curve at its center. This vector is the directional vector and determines the trajectory of the generating curve as it moves through space. The changes in orientation of the directional vector are referenced with respect to a "Frenet frame" of two intersecting planes defined by three orthogonal unit vectors: the tangent vector $t$, the binormal vector $b$, and the normal vector $n$. Unlike the customary Cartesian coordinate system with its fixed three orthogonal axes $x$-$y$-$z$, the Frenet frame and its

236

three orthogonal vectors move with the generating curve, with subsequent positionings of the generating curve in space referenced with respect to previous positions. Okamoto measured changes in orientation and position of the three vectors of the Frenet frame in terms of two parameters: $\kappa$ (curvature) and $\tau$ (torsion). The parameters $r$, $\kappa$, and $\tau$ are algebraically independent of one another, and anisometric morphologies may result in simulations even if the parameters themselves remain constant, due to the effect of size changes (Okamoto 1988a: 41). To ensure that geometrically similar figures have the same parameter values, regardless of size differences, Okamoto scaled his parameters with respect to $r$ to produce the standardized parameters $C$ (standardized curvature) and $T$ (standardized torsion), which we encountered in chapter 4:

$$C = r\kappa \tag{9.12}$$

$$T = r\tau \tag{9.13}$$

Okamoto (1988a,b,c; 1993) was able to produce elegant simulations of highly aberrant shell forms, some of which reflect actual morphologies seen in some heteromorph ammonites, and some of which— produced by the computer—have never been produced by nature. An example of an extremely bizarre growth series, but one that does appear similar to that seen in the actual ammonite *Nipponites*, is given in figure 9.2. Such morphology in ammonites has been called "meandering," like the complex loops and curves seen in a meandering river system, although in this case the meandering is in three dimensions! The complex changes in growth direction of the generating curve necessary to produce the forms seen in figure 9.2 are produced by three *constants* in a growth regulatory model using the moving-reference frame technique. Such a simulation would be virtually impossible using a fixed-reference frame model and would require numerous (and biologically unrealistic) variations in the growth parameters. The theoretical morphologic goal of "parameter minimization" is also met in Okamoto's (1988a) moving-frame methodology, in that quite complex morphologic simulations can be generated with only three variables: $E$, $C$, and $T$, as we saw in chapter 4.

Okamoto (1988c) was able also to simulate what appear to be "morphological saltations" (or "topological instabilities," see Bayer 1978b), major changes in morphology that occur during the ontogeny of the individual, by making minor changes in the parameters of the growth regulatory model that produced figure 9.2. Okamoto argued that the

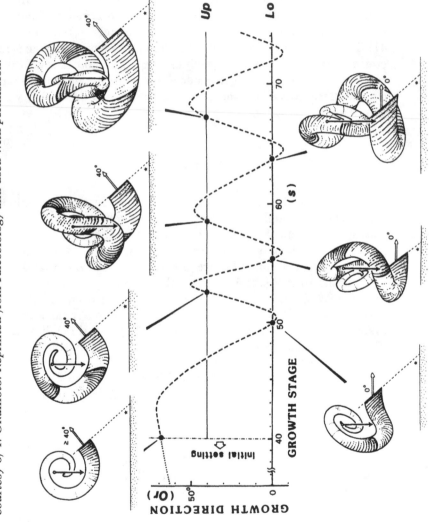

FIGURE 9.2. Simulation of anisometric growth in Okamoto's heteromorph ammonite *Nipponites* (1988c) using a generating-curve model with a moving-reference frame. *Artwork courtesy of T. Okamoto. Reprinted from Paleobiology and used with permission.*

morphologic transition from ammonites like *Eubostrychoceras*, which have open coiling of the whorls but with whorls that still follow a helicospiral, and *Nipponites*, which does not coil helicospirally, occurred abruptly—without any morphologic intermediates—by means of a relatively minor change in the morphogenetic program of *Eubostrychoceras*-like ammonites. Such "morphological saltations" were produced by Okamoto's theoretical morphologic modeling (1988c, 1993), thus are geometrically possible, and the empirical absence of any transition forms in the fossil record further supports his argument.

Ackerly's (1989a) independently derived moving-reference frame model also has three basic parameters but is formulated differently from Okamoto's (1984, 1988a). Okamoto's moving-frame method (1988a) entails two rotations in two separate planes, one involving $\kappa$ and the other involving $\tau$, for each repositioning of the generating curve during growth. The model is also written in terms of differential geometry, that is, in terms of continuous differential equations. Ackerly's moving-frame method uses only one rotation to orient the directional vector of the generating curve during growth and is written in terms of difference equations with incremental steps.

Ackerly (1989a) terms his model "kinematic" because of its emphasis on the stepwise motion of the generating curve explicit in his usage of finite-step difference equations. Similar to Okamoto (1988a), the size of the generating curve in Ackerly's model is defined by a radius $r$ and contains a vector oriented normal to the plane of the generating curve. This vector is termed the "aperture pole," or $P$, and originates from the "centroid" of the generating curve. The centroid of the generating curve is the coordinate origin of the moving-reference frame in Ackerly's model. The generating curve may be any shape, in which the centroid of the generating curve is defined as that point unaffected by dilation of the margin of the generating curve. During growth, the generating curve may "translate," "rotate," and/or "dilate." In simple dilation, the magnitude of $r$ increases with each growth step. In translation, the generating curve moves in space the distance $s$ in a direction specified by the translation axis $T$, which is oriented at the angle $\sigma$ with respect to the aperture pole $P$. In rotation, the generating curve rotates by the angle $\Psi$ about the rotation axis $R$, which is oriented orthogonal to the aperture pole $P$. For a finite growth step of $\Delta t$, growth occurs at the incremental rates $\Delta s/\Delta t$, $\Delta \Psi/\Delta t$, and $\Delta r/\Delta t$. The parameters "translation," "rotation," and "dilation" are algebraically independent of one another, and in any growing system all three might take place, or only one, or a combination of any two. In actual simulations, a sequential order is specified in the model, so that translation occurs first, rotation second, and dilation third (Ackerly 1989a:150).

Ackerly (1989a) reformulated the growth rates in terms of size to define his two parameters "aperture rotation" ($\alpha$) and "aperture dilation" ($\delta$), which were used to create the theoretical morphospace discussed in chapter 4 (see figure 4.7):

$$\alpha = \tan^{-1}\left(\frac{d\Psi}{ds}\right) \qquad (9.14)$$

$$\delta = \tan^{-1}\left(\frac{dr}{ds}\right) \qquad (9.15)$$

In geometric terms, the "aperture rotation" parameter $\alpha$ is the same as the "tangent angle" parameter $\alpha$ (see figure 8.1A and equation 8.3) if growth is isometric. Likewise, the "aperture dilation" parameter $\delta$ is the "apical angle" parameter $\beta$ of Thompson (1942:814–821), which he defines as the angle "which a tangent to the whorls makes with the axis of the shell" (and that many would consider to be one-half the apical angle; Thompson instead terms the value $2\beta$ the "enveloping angle of the cone," rather than the apical angle).

Okamoto's moving-reference frame model (1988a) was modified by Savazzi (1990a) to allow for generating curves that are not circular. This modification has been termed the "template method" and entails inputting an initial set of points that define the perimeter of the generating curve (rather than inputting a single initial radius value). In subsequent growth steps, the coordinates of each of these points are multiplied by a growth factor, so each successive generating curve is a "template" for the next. Savazzi also introduced incremental difference equations, rather than continuous differentials, into Okamoto's model and pointed out that isometric growth might then be obtained by standardizing the increment ratio rather than $\kappa$ and $\tau$ (see equations 9.12 and 9.13). And last, Savazzi introduced a new type of rotation into the model whereby surface twisting of the shell simulation may be obtained by "post-torsion" of the generating curve (the parameter $P$; see Savazzi 1990a:200).

The effect of these modifications can be seen in the elegant computer simulations in figure 9.3. Note that the apertures of the simulated shell forms are not circular and that the surfaces of the shells are twisted or torted in figure 9.3D and F.

Okamoto's (1984, 1988a) and Ackerly's (1989a) moving-frame models and modifications of them (such as Savazzi 1990a,b,c; 1993) are much more versatile than fixed-frame methods and are biologically more realistic. Both also hold much promise for future studies of anisometric growth simulation. Finally, moving-frame theoretical morphogenetic models are easily adaptable to $n$-dimensional theoretical morphospace construction, as we saw in chapter 4.

There is yet another type of model that can be included as a "generating-curve model with moving-reference frames," as it has the essential elements of (1) the generating curve and (2) the specification of a controlling function describing the trajectory of the generating curve through space. These are biomechanical models that simulate the forces or factors determining the shape of the soft-tissue mantle edge that actually secretes the shell increments, for as Morita (1991b:93) argued, the "regulation of shell form can be reduced to the problem of the mechanism controlling the form of the mantle edge at the time of incremental accretion." One such biomechanical model is Hutchinson's (1989)"road-holding" model. In the "road-holding" model, the shape of the preceding whorl is the reference frame determining the positioning of the new whorl, and hence this reference frame moves continually with growth. Hutchinson conceded, however, that such a model could not explain shell form in organisms that do not coil through $2\pi$ revolutions, such as brachiopods and bivalve molluscs.

Another biomechanical mantle-simulation model is the "double membrane structure with internal springs" model, or simply the "DMS-Tube" model, of Morita (1991a,b). The controlling function describing the trajectory of the generating curve, as well as the shape of the generating curve, is formulated in terms of elastic membranes, however, rather than directional vectors. One of the forces determining the shape of the mantle edge in gastropods is the contractive forces produced by the foot. Morita (1993) later proposed that the mode of coiling in gastropods is directly controlled by the positioning of the foot musculature relative to the mantle edge during growth, in a model that he curiously labeled the "dead spiral model." Morita's mechanical models do provide interesting insights into the genesis and placement of structures such as spines and spiral ridges in shells, as well as relationships between generating-curve shapes and the presence or absence of whorl overlap in the shell, but they have yet to be extensively used in theoretical morphology.

For the reader who wishes to experiment further with generating-curve models with moving-reference frames, computer programs using the moving-frame approach are included in Okamoto (1988a:52) and in Savazzi (1990b, 1993).

241

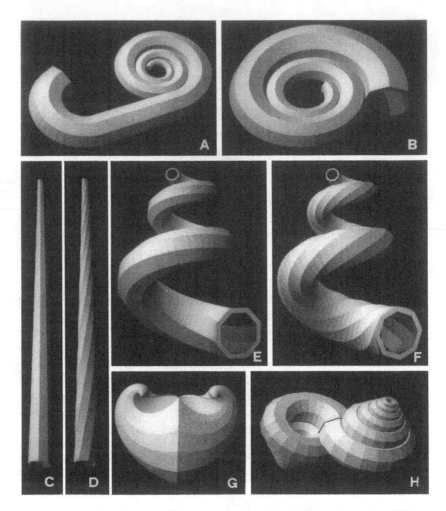

**FIGURE 9.3.** Simulations of accretionary growth in hypothetical molluscs using Okamoto's (1988a) generating-curve model with a moving-reference frame, modified by Savazzi's "template method" (1990a). Figure 9.3A illustrates a hypothetical heteromorph ammonoid, in which the "coiling axis" appears to change position during growth. Figure 9.3B shows a typical planispiral ammonoid, in which the "coiling axis" appears to remain fixed during ontogeny. Figure 9.3C depicts a rectilinear shell that might be produced by a hypothetical scaphopod or orthoconic cephalopod and that has no coiling axis. Figure 9.3D gives a rectilinear shell that has the addition of apertural torsion. In figure 9.3E, the planispiral shell in figure 9.3B has been transformed into an open coiled helicospiral as is found in some gastropods. In figure 9.3F, this helicospiral shell has the addition of "post-torsion," or twisting of the aperture. Two bivalve simulations are given in figures 9.3G and 9.3H: a typical bivalve in which the two valves are mirror

## MULTIVECTOR MODELS WITH NO GENERATING CURVE

In multivector models, the margin of growth is not modeled by a generating curve. Rather, the growth margin is seen as composed of numerous points, which are the termini of numerous independent vectors that together constitute the spirals of the shell. The parameters of growth for each of the vectors may vary greatly across the surface of the growing shell (Bayer 1978b; McGhee 1978a; Savazzi 1987, 1990a,c). In the mathematical example in chapter 8, the transition to a multivector model was made in equations 8.19 through 8.23.

Multivector models offer the freedom to explore a great variety of form changes not possible with generating-curve models. Figure 9.4, for example, shows a series of morphologic deformations in the shape of the aperture of the bivalve mollusc *Barbatia mytiloides*, produced by modifying the growth parameters of the multivectors across the surface of the shell. The visual effect is as if the formerly rigid shell of *Barbatia mytiloides* has been transformed into an elastic membrane (or a rubber shell) that can be stretched, twisted, or compressed into a series of hypothetical morphologies never produced by any individuals of the species. Multivector models can thus be used to explore theoretically the potential variation in form that is possible and also impossible in a particular growing system (Bayer 1978b).

Multivector models are particularly useful for exploring the consequences of anisometric growth. Of course, if the parameters of the vectors are kept the same, isometric growth results. Multivector models have been used to simulate theoretical morphologic features that arise because of allometry and "morphogenetic instabilities" (Bayer 1978b)and to explore major ontogenetic modifications of juvenile morphology, such as bivalves that begin with apertures that lie in a two-dimensional plane in earlier growth stages but that ultimately twist in later growth stages so that the valve aperture now traces a three-dimensional surface (McGhee 1978a; McGhee, Bayer, and Seilacher 1978). Savazzi (1990a) also used a variant of the multivector model to produce his "template method" modification of the generating-curve model with Okamoto's (1988a) moving-reference frames, which was used to produce the elegant simulations seen in figure 9.3.

---

images of each other in figure 9.3G, and an "impossible bivalve" in which the two valves are equal (not reversed) images of each other. *Reprinted from* Biological aspects of theoretical shell morphology *by E. Savazzi from* Lethaia 1990, volume 23, pages 195–212, by permission of Scandinavian University Press. Artwork courtesy of E. Savazzi.

The multivector model was used by Checa (1991) to develop a methodology known as "sectorial expansion analysis" (Checa 1987, 1991; Checa and Aguado 1992). This methodology is designed for the analysis of changes in such geometric features as the curvature and torsion of shells by measuring differential growth among selected sec-

FIGURE 9.4. Computer simulations of hypothetical shell form anisometries using the multivector model. The initial configuration of the vectors is set from measurements taken from the aperture of *Barbatia mytiloides* and then allowed to grow isometrically (top simulation). The effect of introducing differential whorl expansion rates (*W*) and helicospiral components (*H*) in the vectors in the posterior part of the shell is indicated in the subsequent three simulations (indicated by arrows). *From McGhee 1978a.*

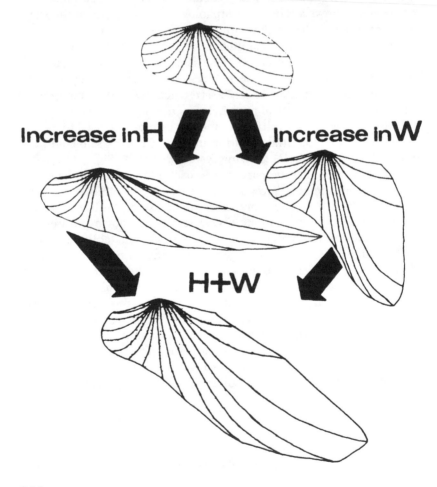

Increase in H     Increase in W

H+W

tors of the shell, using the trace of homologous points along the shell to define the separate helicospirals. The fundamental underlying principle of sectorial expansion analysis is differential growth in different sectors of a shell and is an approach not possible in the generating-curve models with their prespecified geometric figures. Sectorial expansion analysis has been used to apply quantitatively the multivector model to such ontogenetic phenomena as doming and flaring in gastropods (Checa 1991, Checa and Aguado 1992), as illustrated in figure 9.5.

The distortions produced by computer simulation in the shell of *Barbatia mytiloides* (figure 9.4) and the quantitative analysis of differential shell sector expansions or contractions that occur during the ontogeny of *Distorsio reticulata* (figure 9.5) appear superficially similar to the rectilinear grid deformations seen in the method of "coordinate transformation" that Thompson (1945:1026–1090) proposed for comparing related forms. The multivector model is not an empirical system of grid deformations or comparisons, however, but a theoretical model of the process of morphogenesis itself.

For the reader who wishes to experiment further with multivector models, computer programs using the multivector approach are included in Savazzi (1985) and the template approach in Savazzi (1990b, 1993).

## THE EVOLUTION OF THEORETICAL MODELS OF ACCRETIONARY MORPHOGENESIS

A very interesting and innovative examination of the evolution of theoretical morphologic models was conducted by Stone (1996b), using the principles of phylogenetic analysis, or cladistics. Such a procedure is very unusual in that theoretical morphologic models are intellectual constructs and not organisms. The principles of phylogenetic analysis are designed to decipher and replicate the sequence of changes that occur in the morphologic characters of organisms as they evolve with time, in essence, to decipher which character changed first, which character changed second, and so on, in an evolutionary lineage. Characters that are primitive and have not changed from the ancestral condition are termed "plesiomorphic," and a group of species that possess those characters by simple inheritance is said to possess "shared primitive characters," or symplesiomorphies. Characters that have changed with time, or evolved, are termed "derived characters," or apomorphies. A group of species that share a new evolutionary novelty or apomorphic character, inherited from these species latest com-

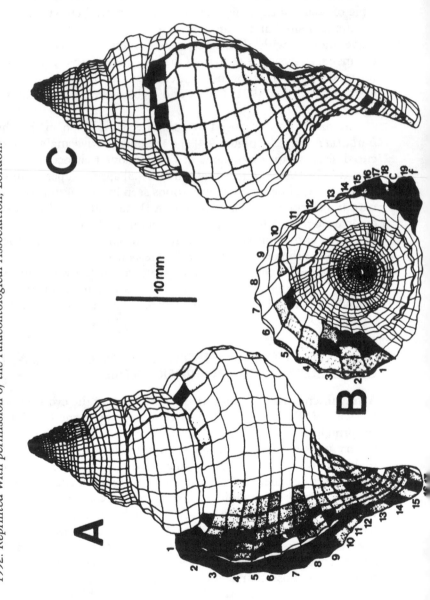

**FIGURE 9.5.** Sectorial-expansion analysis of ontogenetic changes in shell surface form of the gastropod *Distorsio reticulata*. Positive values of the sectorial-expansion rate ($S_c$) indicate anisometric expansion of a shell sector; negative values indicate anisometric contraction; and an $S_c$ of zero indicates isometry. Unshaded regions of the shell surface have an $S_c$ of less than zero; gray regions have rates from zero to 2; and dark shaded regions have rates greater than 2. *From Checa and Aguado 1992. Reprinted with permission of the Palaeontological Association, London.*

mon ancestor, is said to possess a "shared derived character," or syna-pomorph.

The subject of this chapter has been the examination of the theoretical morphologic models of morphogenesis in accretionary growth systems designed in the past three decades. These models are obviously distributed in time, with some younger and some older, but can they be said in any sense to evolve? To evolve, a model must give rise to a descendant model by some modification of the original model, just as one species gives rise to a descendant species by some evolutionary change in the genotype of the original species. Stone (1996b) treated various theoretical morphologic models as "species," and the various features of the models as "morphologic" characters, and then examined the distribution of these "morphologic" characters across the model "species" in a phylogenetic analysis, exactly as one would analyze the distribution of actual morphologic characters across actual species.

One obvious potential difference between a theoretical morphologic model "species" and an actual species is that actual species are separate gene pools and do not hybridize, except in cases of very closely related and recently evolved sibling species. A new model "species," however, could be constructed using the "morphologic" characters of several individual model "species" and not just one. Such a new model "species" would be created by hybridizing ideas from several published sources, and it would not be the descendant through the modification of a single ancestral model "species." The purpose of Stone's analysis was to test this possibility, or as he stated (1996b:917–919), "The dilemma, then, is whether one should treat the models as historical individuals or as natural kinds. If the models are historical individuals, a phylogenetic approach is appropriate, whereas if they are natural kinds, a phenetic approach may be more appropriate."

The distribution across 19 theoretical morphologic model "species" of the 10 "morphologic" characters that Stone (1996b) chose is shown in table 9.1. I have rearranged the ordering of the columns from Stone's original table to match the sequential ordering of the characters as they occur in the cladogram given in figure 9.6. Stone chose Moseley's (1838) and Thompson's (1917) models as outgroups, as these predate the development of computers and computer simulation. He then analyzed the data given in table 9.1 using the computer program Hennig86, which produced two cladograms, one of which is given here for discussion in figure 9.6.

Stone (1996b) concluded that the theoretical morphologic models

**TABLE 9.1** Data matrix used in Stone's 1996b "phylogenetic" analysis of the evolution of theoretical models of accretionary growth systems. The distribution of "morphologic" characters in theoretical model "species," in which the "species" are ranked from the oldest to the youngest, are given below. A zero indicates the absence of that character in the theoretical model (the plesiomorphic condition), and a "Y" indicates its presence (an apomorphic condition). The characters are (1) parameters, (2) computergraphics, (3) algebraic independence of parameters, (4) Frenet frames, (5) variable parameters, (6) anisometric scaling of the aperture, (7) anisometric coiling, (8) generating-curve independence, (9) longitudinal helicospirals, and (10) no reference to an external coordinate system. See the text for discussion.[*]

| Model "Species" | Morphologic Character | | | | | | | | | |
|---|---|---|---|---|---|---|---|---|---|---|
| | *(1)* | *(2)* | *(3)* | *(4)* | *(5)* | *(6)* | *(7)* | *(8)* | *(9)* | *(10)* |
| Moseley 1838 | 0 | 0 | 0 | 0 | 0 | 0 | 0 | 0 | 0 | 0 |
| Thompson 1917 | Y | 0 | 0 | 0 | 0 | 0 | 0 | 0 | 0 | 0 |
| Raup 1961–1966 | Y | Y | 0 | 0 | 0 | 0 | 0 | 0 | 0 | 0 |
| Stasek 1963 | Y | Y | 0 | 0 | 0 | 0 | 0 | 0 | 0 | 0 |
| L–vS–L 1974–1988 | Y | Y | 0 | 0 | 0 | 0 | 0 | 0 | 0 | 0 |
| K–R 1975 | Y | 0 | 0 | 0 | 0 | 0 | 0 | 0 | 0 | 0 |
| B–McG–S 1977–1985 | Y | Y | 0 | 0 | Y | Y | Y | Y | Y | 0 |
| McGhee 1980b | Y | Y | 0 | 0 | Y | Y | 0 | 0 | 0 | 0 |
| Illert 1982–1989 | Y | Y | Y | Y | 0 | 0 | 0 | 0 | 0 | 0 |
| E–C 1983 | Y | 0 | 0 | 0 | 0 | 0 | 0 | 0 | 0 | 0 |
| Okamoto 1984–1988 | Y | Y | 0 | Y | Y | Y | Y | Y | 0 | Y |
| Ackerly 1989a | Y | Y | Y | 0 | Y | Y | Y | Y | 0 | Y |
| Cortie 1989 | Y | Y | Y | 0 | Y | Y | Y | 0 | 0 | 0 |
| Hutchinson 1989 | 0 | 0 | — | 0 | — | 0 | Y | Y | 0 | Y |
| Savazzi 1990a | Y | Y | 0 | Y | Y | Y | Y | 0 | 0 | Y |
| Schindel 1990 | Y | 0 | Y | 0 | Y | Y | Y | Y | 0 | 0 |
| J–T–B 1991 | Y | 0 | Y | 0 | Y | Y | Y | Y | 0 | Y |
| C–A 1992 | Y | 0 | — | 0 | Y | Y | Y | Y | Y | 0 |
| F–M–P 1992 | Y | Y | Y | Y | 0 | 0 | 0 | 0 | 0 | 0 |

[*]Models are designated by their author and date of publication, with the exception of the following "anagenetic" and multiauthored models: L–vS–L for Løvtrup and von Sydow (1974, 1976) and Løvtrup and Løvtrup (1988); K–R for Kohn and Riggs (1975); B–McG–S for Bayer (1977), McGhee (1978a), and Savazzi (1985); E–C for Ekaratne and Crisp (1983); J–T–B for Johnson, Tabachnik, and Bookstein (1991); C–A for Checa and Aguado (1992); and F–M–P for Fowler, Meinhardt, and Prusienkiewicz (1992).

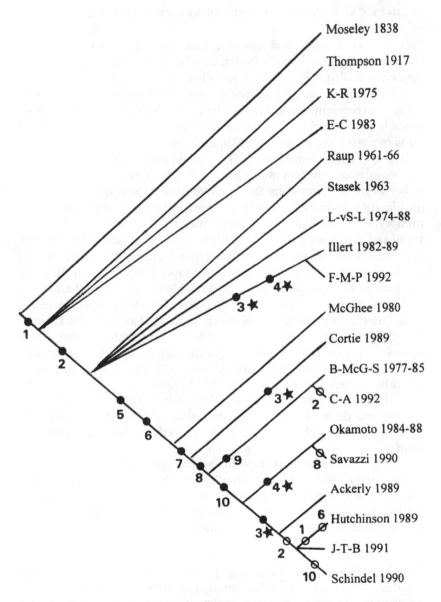

FIGURE 9.6. A "phylogenetic" hypothesis of the "evolutionary" relationships among theoretical morphologic models, as modified from Stone (1996b). Synapomorphic characters are indicated by solid circles and number; convergent characters are indicated by numbers with stars; and evolutionary reversals are indicated by open circles and numbers for characters that have reverted to the plesiomorphic state. The "morphologic" characters are (1) parameters, (2) computer graphics, (3) algebraic independence of parameters, (4) Frenet frames, (5) variable parameters, (6) anisometric

indeed evolved in an ancestor-descendant fashion. Only two of the 10 characters appear to have "convergently" evolved and occur in the cladogram at multiple locations. These characters are "algebraic independence of parameters" (character 3), which "evolved" three separate times independently, and "Frenet frames" (character 4), which arose twice independently. The other eight characters occur in a sequential array of nested synapomorphies (figure 9.6).

The two characters of formally defined "parameters" (character 1), and simulations by "computer graphics" (character 2), are considered to be "key innovations" by Stone (1996b) and to be responsible for the proliferation of models seen in the remainder of the cladogram. The monotypic clades Raup (1961–1966), Stasek (1963), and L–vS–L (1974–1988) and the sister species clade Illert (1982–1989) and F–M–P (1992), constitute a tetrachotomous grouping that Stone (1996b) characterized as "form models" describing size and shape. The remaining 10 models, sharing the synapomorphies of "variable parameters" (character 5) and "anisometric scaling of the aperture" (character 6), constitute a large holophyletic clade that Stone characterized as "growthlike models" describing accretion rather than form. He concluded that a general evolutionary trend exists, in that theoretical morphologic models have become increasingly more realistic, but that this evolutionary trend has taken two routes in the phylogenetic tree (figure 9.6): the evolution of "species" that more realistically simulate the appearance of form and the evolution of "species" that accomplish the same goal by more realistically simulating the growth process that produces form.

I have a few disagreements with the analysis, but they are relatively minor. Stone (1996b) indicates that McGhee's (1980b) model allows for anisometric scaling of the aperture but does not include anisometric coiling. I specifically designed the model in McGhee (1980b) to allow for ontogenetic changes in the parameter $W$, the whorl expansion rate,

scaling of the aperture, (7) anisometric coiling, (8) generating curve independence, (9) longitudinal helicospirals, and (10) no reference to an external coordinate system. The theoretical morphologic models are designated by their author and date of publication, with the exception of the following "anagenetic" and multiauthored models: K–R for Kohn and Riggs (1975); E–C for Ekaratne and Crisp (1983); L–vS–L for Løvtrup and von Sydow (1974, 1976) and Løvtrup and Løvtrup (1988); F–M–P for Fowler, Meinhardt, and Prusinkiewicz (1992); B–McG–S for Bayer (1977), McGhee (1978a), and Savazzi (1985); C–A for Checa and Aguado (1992); and J–T–B for Johnson, Tabachnick, and Bookstein (1991).

which controls the curvature of the spiral in the model (see equations 8.24 through 8.31). Ontogenetic change in $W$ produces spirals that are not isometric or logarithmic; that is, it produces anisometric coiling.

Second, it could be debated whether the highly derived clade of Hutchinson (1989), Johnson, Tabachnick, and Bookstein (1991), and Schindel (1990) is actually a clade of theoretical models of morphogenesis. Note that this clade is defined by the evolutionary reversion in character 2, that is, the absence of computer graphics in the three members of the clade. Note too that the trichotomy of the clade is also defined by the absence, or evolutionary reversion, of synapomorphic characters present at earlier plesions in the cladogram (characters 1 and 6 in Hutchinson 1989 and character 10 in Schindel 1990). One could argue that it was not the intent of the three studies in the clade to produce a new model of morphogenesis by building on previous models and that this might explain the large number of evolutionary reversions seen in the clade. Schindel, for example, was more interested in creating a theoretical morphospace with algebraically independent parameters than with proposing a new model of morphogenesis.

Finally, Okamoto (1996) considers his earlier model (Okamoto 1984) to be a distinctly different class of model from his later "growing-tube" model (Okamoto 1988a) and thus would not place the two models together as a single branch with a unique autapomorphy, as shown in the cladogram (figure 9.6). These disagreements are over detail, however, and do not detract from the essential innovativeness of Stone's approach (1996b). More important is the fact that, unaware of Stone's study, Okamoto (1996) explicitly stated that he considered the model developed later, and independently, by Illert (1987, 1989) to be convergent with his earlier model (Okamoto 1984). And that is exactly the conclusion independently reached by Stone (1996b), as seen in figure 9.6.

# Theoretical Models of Other Aspects of Morphogenesis in Nature

*Computer simulations are devised with the intention of mimicking nature and consequently improving the understanding of natural systems. A successful simulation prompts the inference that the procedures used in the computer program are analogous to those pertaining in nature. Simulation can be regarded as conceptually opposite to analysis: in analysis, observed data from nature are investigated to reveal properties related to the causative process; in simulation, an educated guess at the causative process is used to create artificial data (which may then bear comparison with real data).*
—*Swan (1990b:32)*

*The essence of the method . . . consists in supposing a priori the existence of a differential model underlying the process to be studied and, without knowing explicitly what the model is, deducing from the single assumption of its existence conclusions relating to the nature of the singularities of the process.*
—*Thom (1975:4)*

As the reader perhaps can sense from the previous two chapters, the great majority of morphogenetic models in theoretical morphology have been concerned with accretionary growth systems, found chiefly in the exoskeletons of various invertebrate animal groups. Morphology produced by other growth systems has been investigated using the methodology of theoretical morphology, such as branching growth systems (chapter 3) and discrete growth systems (chapter 6). Some aspects of the morphogenetic modeling of branching and discrete growth systems will be examined in this chapter. In addition, we shall also encounter very interesting morphogenetic models of parts of morphology, rather than the entire form of an organism such as a tree or a foraminiferal skeleton. These partial aspects of form range from fur patterns in mammals to color patterns in butterfly wings and from the dermal scales of sharks to trackways of unknown organisms left on the seafloor in the depths of the oceans.

## BRANCHING MODELS

The literature on models of branching systems, particularly in plants, is enormous and ranges from nonmathematical "architectural" models (Hallé, Oldeman, and Tomlinson 1978) to system models that are derived, of all places, from the field of linguistics ("L-systems"; Prusinkiewicz and Lindenmayer 1990, 1996). In branching models, the fundamental question is "when to branch" and further complexity is then added to this basic level. And that further complexity may be very great indeed! In a review of branching models, Bell (1986) proposed that such models be divided into three types: "blind," "sighted," and "self-regulatory." I shall follow Bell's (1986) classification in the following three sections and then add Lindenmayer systems as an additional type at the end of this section.

### "Blind" Branching Models

In "blind" models, the branching process is controlled by program algorithms without reference to information from the external environment or from within the developing system. In the branching models discussed in chapter 3, Bell (1986) considered the tree model of Honda (1971), the graptolite model of Fortey and Bell (1987), and the bryozoan models of Cheetham, Hayek, and Thomsen (1981) and Cheetham and Hayek (1983; see figure 3.3 in this book) to be examples of "blind" branching models. In each of these models, the complexity of the simulation produced is argued to depend "solely on the complexity of the imposed rules and the duration of the simulation" (Bell 1986:149–150). For example, in Cheetham, Hayek, and Thomsen's model (1980), the number of growing branch tips $G$ at any given time $t$ is determined by the series

$$G_t = G_{t-a} + G_{t-b} \qquad (10.1)$$

where $a$ and $b$ are positive integers, with the value of $a$ greater than or equal to that of $b$. If both $a$ and $b$ have the value of 1, a geometric series in base-2 results (see chapter 3), and if $a$ is equal to 2 but $b$ is equal to 1, the series of numbers follows a Fibonacci sequence (more on Leonardo Pisano, Italian mathematician [circa 1170–1230], aka "Fibonacci," in the section on series later in this chapter).

## "Sighted" Branching Models

"Sighted" models are those in which the branching process is influenced by information from the local surrounding, but without reference to information from within the growing system. Such information from the local surrounding might be the "proximity of other organisms, or parts of the same organism" (Bell 1986:149). Of the models discussed in chapter 3, Bell (1986) considers the bryozoan models of Gardiner and Taylor (1982) and McKinney and Raup (1982; see figures 3.1 and 3.2 of this book) as examples of "sighted" branching models. The effect of "interference" on branching is built into these models, for example, the parameter $XMIN$ in McKinney and Raup's model, which specifies a minimum distance between three adjacent branches that must be exceeded before new branching is initiated. Such models do not necessarily have to be "new" or more recent in time, as Bell (1986) considers the decades-old model of Cohen (1967) also to be an example of a "sighted" model. This work is a very early computer-simulation study of the morphogenesis of the branching growth processes, conducted in spirit of theoretical morphology, although no mention is made of Raup's (1961, 1962, 1966, 1967) work with the computer simulation of accretionary growth systems. That is, Cohen uses the methodology of modern theoretical morphology, as can be seen in the following two statements, that the study is an attempt "to simulate the growth of some biological patterns with the help of a digital computer, using the simplest possible generation rules and the smallest number of parameters in the programmes" (Cohen 1967:246) and that the resulting "simulation programme which incorporates some hypotheses about the generation process of a natural pattern provides a method for an unambiguous rejection of incorrect hypotheses by comparing the natural pattern and its simulation" (Cohen 1967:248). In Cohen's model, the probability of new branching is a function of a computed "local density field" of preexisting branches and their respective distances. Although his was a simple planar two-dimensional model, Cohen was able to simulate hypothetical morphologies that might be found in molds, filamentous algae, mosses, venation in plant leaves, and some aspects of tree morphology (in two dimensions).

## "Self-Regulatory" Branching Models

A third category of branching model is the "self-regulatory" model, which Bell (1986:149) characterizes as a model in which "branch initiation is controlled by the developing simulation itself, using communication via components of the existing framework, whether or not affected by 'environmental' factors." In such a system, the pattern of branching is controlled by internal feedback information from within the growing system, as well as possible external information from the local surrounding. Honda's original model (1971), discussed in chapter 3, is modified to simulate branch "interaction" and the regulation of such interactions by "unequal flow rates" in the later model of Honda, Tomlinson, and Fisher (1981), a model that Bell (1986) considers to be an example of the "self-regulatory" type. The more recent model of Ford, Avery, and Ford (1990), in which branch growth is determined by a balance function of positive and negative internal feedback, is also an example of "self-regulatory" simulation modeling.

## Lindenmayer-System Branching Models

Unlike the algebraic models considered thus far, Lindenmayer-system models simulate branching (and other discrete aspects of morphogenesis, such as cell division) by a system of "rewriting" algorithms. In Lindenmayer-system modeling (usually referred to by practitioners as "L-system modeling"), very complex morphologies are produced by successively "rewriting" parts of an initial simple morphology by using a set of "rewriting rules." L-system modeling was formalized by the biologist Aristid Lindenmayer (1968; see also Sattler 1978), but the concept has roots in the linguistic analysis of grammar syntax. (A brief aside: there is a branch of linguistics also known as "theoretical morphology"). L-system models have extensions from modern fractal analysis to John Conway's mathematical "Game of Life" (Gardner 1970, 1971; see also Dennett 1996:166–176). The best introduction to L-systems is the book *The Algorithmic Beauty of Plants* (Prusinkiewicz and Lindenmayer 1990, 1996), which is filled with fascinating simulations and beautiful color graphics.

In L-system models, one begins with a simple initiator, or "axiom," and in successive growth steps, the axiom is replaced, or "rewritten," following a set of rewriting rules or "productions," eventually resulting in quite complex final morphologies. For example, the process of cell divisions that produce a multicellular filament of the cyanobacterium *Anabaena catenula* was simulated by the following L-system (reviewed in Prusinkiewicz and Lindenmayer 1990, 1996):

$$
\begin{aligned}
&\omega: a_r \\
&p_1: a_r \rightarrow a_l b_r \\
&p_2: a_l \rightarrow b_l a_r \\
&p_3: b_r \rightarrow a_r \\
&p_4: b_l \rightarrow a_l
\end{aligned}
\qquad (10.2)
$$

The letters *a* and *b* represent the size and state of readiness of the cells to divide, and the subscripts *r* and *l* indicate cell polarity, in which daughter cells of type *a* or *b* will be budded to the right or the left. The initiator $\omega$, or "axiom," is a cell of type $a_r$. All subsequent cell divisions follow the "rewriting rules" of the productions $p_1$ to $p_4$; that is, whenever a cell of type $a_r$ occurs in a growth step, it divides to produce two cells of type $a_l$ and $b_r$ in the next growth step, and so on. Starting from a single cell, the multicellular filament is produced from the following cell types:

$$
\begin{aligned}
&a_r \\
&a_l b_r \\
&b_l a_r a_r \\
&a_l a_l b_r a_l b_r \\
&b_l a_r b_l a_r a_r b_l a_r a_r \\
&\cdots
\end{aligned}
$$

An L-system of the form given in equation 10.2 is simply an algorithm of division, bifurcation, or branching types. More complicated graphic interpretations of the rewriting rules are needed to simulate complex branching systems in three dimensions. The literature outlining various graphic coding models is abundant and is reviewed in Prusinkiewicz and Lindenmayer (1990, 1996), who themselves favor the "turtle" interpretation model (after Abelson and DiSessa 1982). The "turtle" has an initial position and a directional heading, called its "state." In three dimensions, the state of the turtle is defined by three orthogonal unit vectors that indicate the turtle's "heading," the direction to the "left," and the direction "up." Future motion of the turtle through space with time is given in a series of L-system productions, which specify at each step a series of motions, turns, or rotations, each of which then is "rewritten" in the next step. By using such a procedure, what appear to be extremely complex fractal curves or tree-branching systems can be produced with simple L-systems (see Prusinkiewicz and Lindenmayer 1990, 1996).

L-systems can be either deterministic or stochastic, context free ("blind" in the sense of Bell 1986) or context sensitive ("self-regulatory" in the sense of Bell 1986). Equations of growth functions can be rewritten in terms of L-systems. For example, the L-system

$$\omega: a$$
$$p_1: a \rightarrow ab \qquad (10.3)$$
$$p_2: b \rightarrow a$$

will produce a string of letters in which the numbers of the letter $a$ follows the Fibonacci sequence. Prusinkiewicz and Lindenmayer (1990, 1996) also rewrote such diverse preexisting models as the tree-generating branching model of Honda (1971) and the sunflower-head phyllotactic-spiral model of Vogel (1979) in terms of L-systems. In chapter 3 (see figure 3.5), we saw these elegant monopodial treelike computer simulations, produced by L-systems. Even more realistic plantlike forms, entirely hypothetical and computer produced by L-systems, are illustrated in figure 10.1.

Although very complex morphologies may be generated by L-systems with small numbers of productions, or rewriting rules, the "specification of L-systems is not a trivial task" (Prusinkiewicz and Lindenmayer 1990:63). And even though L-systems allow very elegant morphogenetic modeling and computer simulation of both existent and nonexistent forms in nature, it is difficult to parameterize such a system for theoretical morphospace construction.

## SERIES AND SEQUENCE MODELS

In previous considerations of both branching systems (chapter 3, and thus far in this chapter) and discrete growth systems (chapter 6), we encountered models that employ series. For example, in a simple bifurcation system in which one branch divides into two, and then each of these two themselves divide into two new branches, the total number of branches $N$ in the system will follow a geometric series in base 2:

$$N = 2^0 + 2^1 + 2^2 + 2^3 + \cdots + 2^n \qquad (10.4)$$

A geometric series is also seen in the parameters of Berger's model (1969) for simulating the discrete addition of chambers that take place

**FIGURE 10.1.** Examples of the computer simulation of plantlike structures using L-systems. *From Prusinkiewicz and Lindenmayer (1990, 1996). Copyright ©1990 by Springer-Verlag and Przemyslaw Prusinkiewicz and used with permission.*

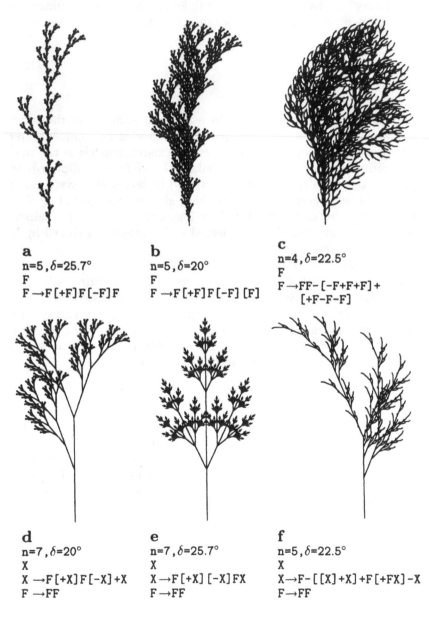

**a**
n=5,δ=25.7°
F
F →F[+F]F[-F]F

**b**
n=5,δ=20°
F
F →F[+F]F[-F][F]

**c**
n=4,δ=22.5°
F
F →FF-[-F+F+F]+
    [+F-F-F]

**d**
n=7,δ=20°
X
X →F[+X]F[-X]+X
F →FF

**e**
n=7,δ=25.7°
X
X →F[+X][-X]FX
F →FF

**f**
n=5,δ=22.5°
X
X →F-[[X]+X]+F[+FX]-X
F →FF

in the growth of a foraminiferal skeleton (see chapter 6). A general expression for the geometric series is

$$y = a + aq + aq^2 + aq^3 + \cdots + aq^n \qquad (10.5)$$

If we take $a$ as equal to 1, and $q$ as equal to 1.1, then the terms of equation 10.5 are the geometric sequence 1, 1.1, 1.21, 1.331, . . ., which we saw in the calculation of Berger's (1969) *q-ratio* parameter for foraminifera (equation 6.1; likewise, the terms for the *a-angle* parameter may be obtained by setting $a$ as equal to 0.4 and $q$ as equal to 1.1 in equation 10.5).

Undoubtedly the most famous mathematic sequence is that of the Italian mathematician Leonardo Pisano ("Fibonacci") who, in addition to introducing into Europe the numerals we today call "arabic," was concerned with breeding rabbits in the year 1202. The solution to the problem of "how many pairs of rabbits do I have?" with each successive reproductive cycle turns out to be the "Fibonacci sequence," in which each number in the sequence is the sum of the two previous numbers: 1, 1, 2, 3, 5, 8, 13, 21, 34, 55, 89, 144, 233, . . ., and so on, where

$$a_{n+1} = a_{n-1} + a_n \qquad (10.6)$$

if $n$ are integers of value 2 or larger (the "recursion formula"; see Batschelet 1974. For a formulation of the Fibonacci sequence in L-systems, see equation 10.3).

In morphology, the Fibonacci sequence figures most prominently in the great phyllotaxis debate. The term "phyllotaxis" literally means "leaf arrangement," and the analysis of the geometric arrangement of leaves in plants has been the focus of attention dating back at least to Leonardo da Vinci, if not to the ancient Greek mathematicians (Thompson 1942:912–933). In short, leaf arrangements in plants follow a Fibonacci sequence. And it is not just leaves; the arrangement of stalks in a celery plant, of florets on a pineapple, leaf scars on palm trees, petals in flowers, scales in pinecones, and seeds in sunflower heads also follows a Fibonacci sequence. Indeed, the Fibonacci sequence appears over and over in the plant kingdom, an observation that many have taken to indicate the action of a fundamental law of growth, on one hand, to the mere "arithmetical playing with ideas" (J. Sachs, quoted in Thompson 1942:912), on the other. Phyllotaxis has been described, in awe, as the "bugbear of botany" (E. J. Corner, quoted in Stevens 1974:133).

What is the debate all about? Consider a helicospiral arrangement of leaves around a branch, or subbranches around a main branch, or branches around a main tree trunk. Take the lowest leaf on the branch as the first leaf, the next leaf higher up along the branch as the second leaf, and so on. The second leaf on the branch is usually oriented at an angle to the first leaf; that is, the second leaf is not directly above the position of the first leaf on the branch. Farther along the branch, however, we usually encounter a leaf, say the $n$th leaf, which is directly above the position of the first leaf. We can then use the first leaf and the $n$th leaf to define a "phyllotactic period," a period that has two descriptive components: the number of leaves in the period (in this case, $n$) and the number of whorls or windings that the leaves make around the branch in the period, which we can designate as $w$. Empirical examination of enormous numbers of plants reveals that both $n$ and $w$ are numbers in the Fibonacci sequence. For example, $w$ is 1 and $n$ is 2 in the horizontal twigs of the elm, $w$ is 8 and $n$ is 21 in the scales in the cones of spruce and fir, but $w$ is 13 and $n$ is 34 in the scales of pinecones of *Pinus* (Schips 1922). In all cases, the values of $w$ and $n$ are two numbers in the Fibonacci sequence that are not immediately adjacent to each other but are separated by a single number within the sequence. This interesting relationship is formalized as the Schimper and Braun "phyllotactic fraction" sequence (Jean 1984)

$$1/2, \ 2/5, \ 3/8, \ 5/13, \ 8/21, \ \ldots, \ a_n/a_{n+2} \qquad (10.7)$$

where $a$ are the numbers in the Fibonacci sequence. The numerator represents the number of whorls or revolutions around the plant's branch or trunk, and the denominator is the number of leaves, twigs, branches, and so on. Considering trees alone, the phyllotactic fraction 1/2 is found in elms, lindens, and mulberry trees; the phyllotactic fraction 1/3 is found in alders, beeches, and birches; the phyllotactic fraction 2/5 is found in plums, oaks, and cherry trees; the phyllotactic fraction 3/8 is found in poplars, sycamores, and pear trees; the phyllotactic fraction 5/13 is found in willows, almonds, and white pine trees; and so on (Jean 1984).

Fibonacci fractions are also found in compound spiral arrangements in plants, such as seeds in sunflower heads or scales in pinecones, in which the numerator and the denominator of the fraction are now the number of clockwise (dextral) spirals versus the number of counterclockwise (sinistral) spirals, respectively. However, the sequence of the numerator and denominator values are now successive terms in the Fibonacci sequence

$$1/1, \ 1/2, \ 2/3, \ 3/5, \ 5/8, \ 8/13, \ 13/21, \ \ldots, \ a_n/a_{n+1} \qquad (10.8)$$

unlike those in sequence 10.7. For example, Stevens (1974) illustrates a celery plant in which the spiral arrangement of stalks is in a fraction of 1/2, a pineapple with florets spirally arranged in a fraction of 8/13, a daisy with a fraction of 21/34, and he mentions the reported existence of a monster sunflower with a spiral arrangement of seeds having a fraction of 144/233!

In addition to Fibonacci or phyllotactic periods and fractions, we have the "Fibonacci angle" of 137.5°. The angular relationship between the intersection points of the dextral and sinistral spirals in a sunflower head, measured from the center of the flower to its perimeter, all occur in arc segments 137.5° away from one another. The addition of many leaves or twigs around a branch also follow the Fibonacci angle, which leads directly to the placement of leaves and twigs above one another along the branch in a Fibonacci sequence.

Phyllotactic periods and Fibonacci angles in plants have been argued by many to be the simple result of any growth process in which new growth units (leaves, seeds, and so on) are fit into the growing structure where the most room or open space occurs between previously created units (see Stevens 1974, Vogel 1979). Entire books have been written on the mathematics of phyllotaxis (e.g., Jean 1984, 1994), mathematics that incorporate aspects of branching models (considered previously in this chapter) into physicochemical models (which follow later).

Where does theoretical morphology fit into the phyllotactic debate? Theoretical morphology may contribute through branching models or physicochemical models, but all the Fibonacci sequence examples given in the previous discussion are mathematic characterizations of form; that is, they are morphometrics, not theoretical morphology.

In terms of series of sequence models themselves, a theoretical morphologic approach might be to attempt to simulate plant forms using sequences *that are not a* Fibonacci sequence. That is, we could attempt to create nonexistent plant forms, since it appears that so many existent plant forms *are* characterized by a Fibonacci sequence, and to consider why these possible but nonexistent plant forms do not occur in nature. The study of receptaculitid form of Gould and Katz (1975), discussed in chapter 6, represents a step in this research direction. By far the most innovative work in the theoretical examination of plant form that both does and does not conform to Fibonacci sequences is that of Niklas (1997b), whose research we encountered in chapters 3 and 7. For example, in figure 10.2, the effect of leaf shape

FIGURE 10.2. Niklas's "phyllotactic morphospace" (1997b). Multiple computer-simulated adaptive walks, maximizing light interception efficiency (*E*), through the morphospace converge on a leaf divergence angle (θ) of 137.5°, which is the same number converged on by the series of the higher Fibonacci fractions (fractional positions given at the top of the figure). Narrow, slender leaves arranged in Fibonacci sequences have the highest efficiencies of light interception (see the right margin of the figure); leaves with a nearly circular outline have much lower efficiencies, even arranged in Fibonacci sequences, owing to overlapping leaf outlines leading to shading. *Artwork courtesy of K. J. Niklas. From Niklas 1997b. Copyright © 1997 by the University of Chicago and used with permission.*

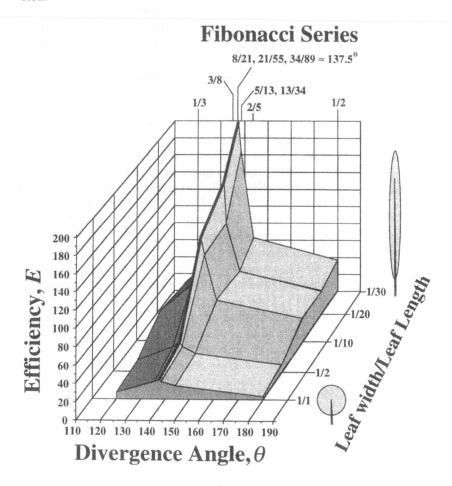

in plants is demonstrated to have a major effect on efficiency of light interception, in addition to leaf arrangement in Fibonacci sequences. Hypothetical plants with circular leaves may have leaf arrangements that deviate significantly from Fibonacci sequences, as leaf divergence angles of 137.5° convey only a slight increase in light interception efficiency (figure 10.2).

Another popular series in morphologic analysis is the Fourier series, a trigonometric polynomial of the general form

$$y = a_0 + a_1 \cos \theta t + a_2 \cos 2\theta t + \cdots + a_n \cos n\theta t$$
$$+ b_1 \sin \theta t + b_2 \sin 2\theta t + \cdots + b_n \sin n\theta t \qquad (10.9)$$

where $a$ and $b$ are coefficients, $\theta$ is angular frequency, and $n$ is the harmonic number. When $n$ tends to infinity, the polynomial is known as a "Fourier series," and "fitting periodic data by a trigonometric polynomial is called *harmonic analysis* or *Fourier analysis*" Batschelet (1974:116, italics Batschelet's). The degree of accuracy of the fit is a function of the number of terms in the polynomial, which is a function of the number of harmonics $n$.

The key words here are "fitting" and "data." In morphologic analyses, Fourier series are most often fit to outline data taken from existent form, and the series coefficients and harmonics are used as a mathematic characterization of that form. Such an analytic procedure is morphometrics, and not theoretical morphology. However, Chapman, Rasskin-Gutman, and Weishampel (1996:66) state that "certain morphometric methods, such as Fourier analysis ... can be used to develop both theoretical and empirical morphospaces." An example in which a Fourier series was used to create a hybrid morphospace was given in chapter 2 (see figure 2.6), from the work of Waters (1977).

## PHYSICOCHEMICAL DEVELOPMENTAL MODELS

*If living organisms are to be envisaged non-vitalistically, as physical systems of reacting chemicals, the kinds of possible preconception can be classified according to the main divisions of physicochemical theory. These are three: structure, equilibrium and kinetics.*

Harrison (1987:370)

The early physicochemical models of development actually predate Raup's initial papers in theoretical morphology by more than four decades (Thompson 1917). However, physicochemical developmental models are models of morphogenesis, and they certainly are theoretical, so I include them here under the larger umbrella concept of modern theoretical morphology. The more recent generation of physicochemical developmental models extensively use computers to simulate patterns found in nature and so clearly fit the definition of theoretical morphologic modeling.

As Harrison (1987) pointed out, there are three main types of physicochemical developmental models: structural, equilibrium, and kinetic (table 10.1). Structural, or "self-assembly" models generally model the totality or shape of existent morphology as a function of

TABLE 10.1. The three classes of physicochemical developmental models and the major divisions of models in each class, with examples of form simulation studies.*

| Model Type | Examples |
| --- | --- |
| I.   Kinetic models | |
| 1. Reaction-diffusion models | ■ color patterns in sea shells (Meinhardt 1995) |
| | ■ color patterns in butterfly wings (Nijhout 1991) |
| | ■ graptolite zooid form (Urbanek and Uchmanski 1990) |
| | ■ color patterns in mammal fur (Bard 1981) |
| 2. Mechano-chemical models | ■ cell rearrangement and shape (Weliky and Oster 1990) |
| | ■ vertebrate limb development (Oster, Murray, and Harris 1983; Oster, Murray, and Maini 1985) |
| | ■ epithelial invagination (Odell et al. 1981) |
| | ■ vertebrate nervous system development (Jacobson 1980) |
| 3. Intercellular-signal models | ■ color patterns in sea shells (Ermentrout, Campbell, and Oster 1986) |
| | ■ neural connections development (Willshaw and von der Malsburg 1976) |
| 4. Self-electrophoretic models | ■ amphibian limb regeneration (Jaffe 1982) |
| II.  Equilibrium models | |
| 1. Membrane (Pneu) models | ■ pneu models in general (Seilacher 1991) |
| | ■ echinoid skeletal form (Dafni 1986, Ellers 1993) |
| | ■ echinoid skeletal form (Thompson 1917, 1942) |
| 2. Free-energy minimization models | ■ organogenesis in plants (Green and Poethig 1982) |
| | ■ cell division patterns (Thompson 1917, 1942) |
| 3. Cell-as-molecule phase-separation models | ■ embryonic cell arrangement (Steinberg 1970) |
| III. Structural (self-assembly) models | ■ spicule biocrystallization in sea urchins (Inoue 1982) |

*Updated and expanded from Harrison's classification 1987.

the static geometric properties of the many smaller units that fit together to constitute the larger whole. Such modeling has been particularly successful in simulating form at the molecular level and has also been extended to larger scales, such as that of the form of viruses and certain skeletal elements of larger organisms (Inoué 1982).

Equilibrium models generally model morphology as the result of processes tending toward an equilibrium or stable state. "Soap bubble" models of form (Thompson 1917, 1942), and other models in which form is the product of the minimization of free energy, the minimization of surface tension, and so on, are equilibrium models (see the examples in Stevens 1974). We have encountered equilibrium models in chapter 6 with the "membrane" or "pneu" models of echinoderm skeletal morphology (see table 10.1).

Kinetic models are by far the most popular of the three types of physicochemical developmental models (table 10.1). Unlike equilibrium models, kinetic models generally model form "as being generated by movement *away from* equilibirium, explicable therefore only in terms of rates of chemical reactions and transport processes" (Harrison 1987:370, italics Harrison's). Of the kinetic models, reaction-diffusion models have been the most extensively utilized by theoreticians engaged in the computer simulation of morphology. Dating back to Turing (1952), reaction-diffusion models assume the existence of hypothetical substances, termed "morphogens," which diffuse through the developing organism and either trigger cell activity or inhibit cell activity. For example, Urbanek and Uchmański (1990) modeled the morphology of uniaxiate graptolite colonies in terms of the diffusion of a morphogen from the sicula outward into the colony. They successfully modeled the observed pattern of size differences in graptolite thecae in terms of the diffusion gradients of this hypothetical morphogen, particularly the observed distal increase in thecal size away from the sicula.

The most widely utilized reaction-diffusion models have two or more morphogens. The two morphogen models assume the existence of two hypothetical substances in the developing organism: an "activator" morphogen that is autocatalytic (i.e., the substance promotes its own production) and an "inhibitor" morphogen that is antagonistic to the effect of the activator morphogen or to the activator morphogen itself. Activator-inhibitor interactive models have been extensively used in the computer simulation of what appear to be extremely complex color patterns, such as those found in mammal fur (Bard 1981), mollusc shells (Meinhardt 1984, 1995; Meinhardt and Klingler 1987), and butterfly wings (Nijhout 1991; see also the general review of color

265

patterns in Seilacher 1991). In an early paper, Bard (1981) successfully simulated complex patterns in mammal fur such as spots (like those found in young deer), complex spots (like those in leopards), reticulated patterns (giraffes), rings (as in raccoons' tails or the tails of domestic shorthair varieties of *Felis catus*), horizontal stripes (like those in the wild "fishing cat" *Prionailurus viverrinus*), and vertical stripes (zebras) using a simple reaction-diffusion model. Meinhardt (1984) used reaction-diffusion models to simulate aspects of the development of segmentation and appendage formation in arthropods, before turning his attention to the complex pattern of colors found in the many mollusc shells that are beautifully illustrated in Meinhardt's (1995) book *The Algorithmic Beauty of Sea Shells*.

Harrison (1987:370) stated that the "language" of reaction-diffusion theory "is necessarily that of the partial differential equation, with both position and time as variables." Why are reaction-diffusion models written in terms of partial differential equations? Good question. Let us consider the diffusion of some substance, say a gas or a liquid, through x-y-z space filled with another substance (say a solvent) with time t. To make things simple, let us consider the spread of this gas or liquid in just one direction (say from one end of a pipe filled with a solvent to the other end), the x-direction. The concentration of the diffusing substance, C, is thus some function of distance in the x-direction, x, and the length of time, t, that the substance has been diffusing, that is, some unknown function $C(x, t)$. We assume that the unknown function $C(x, t)$ is differentiable with respect to x and t, both of which are variables. So what to do? We can hold t constant for the moment and *partially* differentiate $C(x, t)$ with respect to x. The resulting partial derivative $(\partial C/\partial x)$ is known as the "gradient of the concentration" of the diffusing substance. It can then be shown that the rate of change of the concentration of the diffusing substance with time is (see any good mathematics text, such as Batschelet 1974, for the derivation):

$$\frac{\partial C}{\partial t} = D \frac{\partial^2 C}{\partial x^2} \tag{10.10}$$

where D is known as the "diffusion constant." The rate of change of the concentration of the diffusing substance is thus a function of its own gradient of concentration, and the partial differential equation 10.10 is known as the "diffusion equation in one dimension" (Batschelet 1974).

To further illustrate the "language" of reaction-diffusion models, let us consider the elementary conditions for a two-morphogen model, one containing an activator and an inhibitor, as discussed in Meinhardt's (1995) general review. The autocatalytic morphogen, or activator, is designated as the function $a(x, t)$, and the antagonist morphogen, or inhibitor, as the function $b(x, t)$. A basic model for an activator-inhibitor interactive system is given by two equations (Meinhardt 1995:equations 2.1.a and 2.1.b):

$$\frac{\partial a}{\partial t} = s\left(\frac{a^2}{b} + b_a\right) - r_a a + D_a \frac{\partial^2 a}{\partial x^2} \qquad (10.11)$$

$$\frac{\partial b}{\partial t} = sa^2 - r_b b + D_b \frac{\partial^2 b}{\partial x^2} + b_b \qquad (10.12)$$

where although much more complicated, the similarity of equations 10.11 and 10.12 to the "diffusion equation in one dimension" (equation 10.10) can easily be seen. In equations 10.11 and 10.12, we have the variables of position $(x)$ and time $(t)$, and two diffusion constants, one for the activator $(D_a)$ and one for the inhibitor $(D_b)$. In addition, we have the variables $r_a$ and $r_b$, which are the "decay rates" of the activator and inhibitor, respectively, and $b_a$ and $b_b$, which are measures of the "basic production" of the activator and inhibitor, respectively. Last, the variable $s$ is the "source density," which is a measure of the ability of the cells in the model to perform the autocatalysis.

The rate of change of the activator with time is a function of three terms (see equation 10.11). The first term is the "production rate" of the activator, in which the activator has a nonlinear autocatalytic effect $(a^2)$ and which is further a function of the inhibitory effect of the antagonist $(b)$ and the ability of the model cells to autocatalyze $(s)$. The second term in the equation is the "rate of removal" of the activator, which is a function of the decay rate of the activator $(r_a)$ and its own concentration. The third term is the "exchange by diffusion" of molecules of the activator, similar to the expression we saw earlier in the "diffusion equation in one dimension" (equation 10.10). The equation for the inhibitor (equation 10.12) is likewise read in a similar fashion (Meinhardt 1995).

The model given in equations 10.11 and 10.12 is a simple two-morphogen model of activator-inhibitor interaction. Obviously, the reaction-diffusion model can be made more complex, and Meinhardt (1995) outlined many of those additional complexities, such as modeling the effect of saturation of the autocatalytic process, antagonism of the autocatalytic process by the additional factor of the depletion of a substrate substance needed for autocatalysis, inhibition due to direct activator destruction, and, alternatively, indirect autocatalysis by inhibition of the inhibitor. In addition, the model can be made a three-morphogen model by the addition of a second inhibitor or by adding an inhibitor under the control of a hormonelike substance that varies with time.

Meinhardt and Klinger (1987) and Meinhardt (1995) were successful in the computer simulation of many very complex color patterns found in mollusc shells with relatively simple reaction-diffusion models (see figure 10.3). Complex oblique lines of color in shells can be produced by "traveling waves" of activation through the pigment-producing cells in the mantle of the mollusc; complex branches and crosses can be produced by "shifting from oscillatory to steady-state

FIGURE 10.3 Computer simulation of shell pigmentation patterns (background plot), compared with those actually seen in the gastropod *Oliva porphyria* (shell in lower right), produced by Meinhardt's reaction-diffusion models (1995). *Artwork courtesy of H. Meinhardt. Copyright © 1995 by Springer-Verlag and used with permission.*

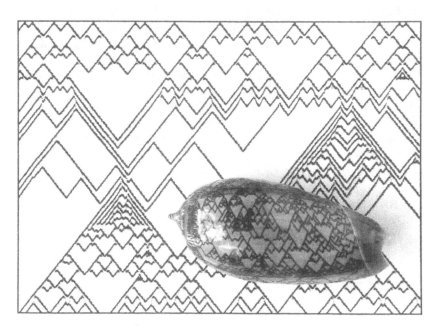

modes" of activation; more complex chessboard or meshworklike patterns can be produced by shifting to three-morphogen models with one autocatalytic activator morphogen that is antagonized by two separate inhibitor morphogens; and so on.

All the reaction-diffusion models assume the existence of hypothetical morphogens, and this common feature brings us to the first of a series of perplexing problems with physicochemical developmental models in general. The first problem is specific to reaction-diffusion models: what are these morphogens? What specific diffusing chemical "activates" a pigment cell to secrete color, for example, and what specific diffusing chemical "inhibits" that same cell to cease secreting pigment? In general, it is not known, and morphogens remain hypothetical. They are the creation of theoreticians, and their existence can be demonstrated only by experimentalists. Unfortunately, the "reluctance of morphogens to stand up and be chemically identified has led to a corresponding reluctance on the part of experimental biologists to treat reaction-diffusion as a promising field and give it extensive experimental tests" (Harrison 1987:381).

A second problem with reaction-diffusion models is that different models can produce very similar, if not identical, morphologic results. For example, Meinhardt and Klinger (1987:67) admit that the "activator-substrate model . . . and the activator-inhibitor model . . . lead to very similar patterns. Therefore, it is not possible to distinguish which mechanism is involved on the basis of shell pattern. The selection of one or the other mechanism for the simulation is more or less arbitrary."

This second problem, that of different models producing the same result, is actually a problem of kinetic models in general, and not just of the reaction-diffusion type. As we said earlier, the existence of morphogens in the mantle of molluscs remains hypothetical. The existence of nerves does not! Nerves in molluscs are an empirical observation; they exist. Ermentrout, Campbell, and Oster (1986) maintain that morphogens are not necessary to explain or simulate the color patterns found in mollusc shells. Instead, they proposed a intercellular-signal kinetic model (see table 10.1) for color pattern production, one that uses the molluscs' existing nervous system. In the neurokinetic model, cells in the mantle are stimulated to produce pigment directly by the molluscs' central nervous system. Cessation of pigment production within the cell is modeled as due to the buildup of an "inhibitory substance" within the cell itself. Thus the neurokinetic model is still an "activator-inhibitor" model, but one that does not depend on diffusing morphogens within the organism. Because reaction-diffusion models depend on diffusion of morphogens from local regions of concentration or production, such models were termed "nearest neighbor" models by Ermentrout, Campbell, and Oster (1986), who argue, "This model differs from previous models in at least one important aspect: it depends on the 'nonlocal' property of nerve nets. That is, because innervations may connect secreting cells that are not nearest neighbors, the possibility of cooperative, long-range interactions is present" (Ermentrout, Campbell, and Oster 1986:370).

In the neurokinetic, or intercellular-signal, model a "neural stimulation function," $S[P_t(x)]$, is specified, a function that contains both "excitatory" and "inhibitory" components that affect pigment secretion, $P$, in cells at position $x$ and at time increments $t$ (taken to be a day in the model). The model is set up so that pigmentation at each time increment $t + 1$ is a function of the pattern of pigmentation at the previous time increment $t$, that is, that "each day's pattern of excitation is stimulated by 'tasting' the previous day's pigment pattern" (Ermentrout, Campbell, and Oster 1986:384). Two functions sum the average pattern of excitation (equation 10.13) and inhibition (equation 10.14) around the "domain" ($\Omega$) of the neurons in the mantle

$$E_{t+1}(x) = \int_\Omega W_E(x' - x) P_t(x') \, dx' \qquad (10.13)$$

$$I_{t+1}(x) = \int_\Omega W_I(x' - x) P_t(x') \, dx' \qquad (10.14)$$

where the connectivity functions, or "kernels," $W_E(x' - x)$ and $W_I(x' - x)$ define the "connectivity of the of the mantle neuron population" by specifying the excitatory and inhibitory effect of neural contact between cells at positions $x'$ and $x$.

A mollusc or brachiopod shell is a four-dimensional object. It exists in x-y-z space but also contains the complete record of the accretionary growth of the organism in time $t$. If Ermentrout, Campbell, and Oster's model (1986) is correct, the color pattern preserved in the shell is also a direct record of the neural activity in the organism's mantle; that is, the shell is an "electroencephalogram" as well!

Using the neurokinetic model, Ermentrout, Campbell, and Oster (1986) successfully simulated such common mollusc color patterns as vertical stripes, horizontal stripes, diagonal stripes, divaricate patterns, wavy stripes, "streams," and interactive checks and "tents." Ermentrout, Campbell, and Oster (1986:388) argued for their model with observations such as "it is not clear how the diffusion-reaction model handles the problem of pattern alignment between episodes of shell secretion, whereas this is intrinsic to the neural model." They also concluded that

one can model the phenomenon of local activation and lateral inhibition characteristic of neural nets in a variety of ways. . . . Therefore, we are left with the disappointing conclusion that it may be quite difficult to infer mechanism from pattern alone, because several quite distinct cellular mechanisms can produce identical patterns. (Ermentrout, Campbell, and Oster 1986:383; for yet another type of computer simulation of shell ornamentation patterns, see Hayami and Okamoto 1986)

In both the reaction-diffusion and intercellular-signal kinetic models of color patterns in seashells (table 10.1 classification), the activator-inhibitor interactive system is essential to the production of morphologic pattern, yet seemingly identical resultant patterns can be produced by very different kinetic systems. For more examples of this fundamental ambiguity, let us consider some that deal with morphologic patterns other than pigmentation. Reif (1980) called for "inhibitory fields" in his model of chondrichthyan scale morphogenesis, and Weishampel (1991) used "inhibition zones" in his model of vertebrate tooth morphogenesis. Both models produce simulations that match real shark dermal skeletons and vertebrate teeth. Yet Weishampel (1991:301) went further, explicitly acknowledging the ambiguity of the fundamental process producing the inhibition zones necessary for the model:

The zone of inhibition can arise along two different courses. It may be produced by some sort of regulatory chemical substance that diffuses through extra-cellular matrix (chemical prepatterning models; Meinhardt 1982; Meinhardt and Klinger 1987). Or it may result from the local recruitment of cells around the initiator tooth (mechanochemical models; Oster et al. 1983, 1985).

That is, Weishampel (1991) considered reaction-diffusion models or mechanochemical models (table 10.1 classification) equally likely for the production of inhibition zones in his model. Yet reaction-diffusion, mechanochemical, and intercellular-signal models make very different fundamental assumptions concerning morphogenesis, and it is indeed a "disappointing conclusion" that theoretical simula-

271

tion modeling seems to offer no clear indicator of the actual morphogenetic process involved and that the ultimate demonstration may perhaps be accomplished only by experimentation.

Theoreticians generally seek simplifying solutions. Harrison (1987) points out that there are three distinctly different divisions of physicochemical developmental theory, as discussed earlier in this section of the chapter, yet certain morphologies may exhibit aspects of all three! Even the morphogenesis of inorganic form, such as that found in snowflakes, can be argued to have self-assembly, equilibrium, and kinetic aspects (Harrison 1987). Erickson (1982) outlines self-assembly, equilibrium, and kinetic aspects in plant morphogenesis, depending on scale. Bard's (1981:363) model of mammalian fur patterns is kinetic (reaction-diffusion), yet later study has suggested that equilibrium (cell-adhesion) factors might be involved as well; thus the generation of mammalian fur patterns may exhibit both nonequilibrium and equilibrium aspects (reviewed in Harrison 1987).

A final problem with physicochemical developmental models may simply be that nature is complex. And even complexity itself may not be unambiguous to model. For example, extremely complex morphologic patterns can also be created with "cellular automata," mathematical models with very simple component parts but very intricate overall behavior (Wolfram 1984a). The complex patterns generated by a cellular automaton are the result of the cooperative effect of its many identical components. Cellular automata simulate complexity by using very large numbers of discrete identical components, components that evolve in discrete steps. They are thus very different in concept from complexity simulation in models that use differential, or partial differential, equations with a comparably small number of variables, variables that change continuously.

Wolfram (1984a,b,c) argued that in addition to many other aspects of complexity in nature, the complexity patterns produced by activator-inhibitor physicochemical developmental models may be produced equally well by cellular automata. And indeed, Wolfram (1984a,c) used cellular automata successfully to simulate spatial patterns that are very similar to the pigmentation patterns found in many mollusc shells and very similar to those produced by the reaction-diffusion model of Meinhardt (1995) and the intercellular-signal kinetic model of Ermentrout, Campbell, and Oster (1986).

## THEORETICAL MORPHOLOGIC MODELS OF BEHAVIOR?

When thinking of theoretical models of behavior—most often game theory, models of how altruism could arise in nature, models of kin selection, and so on—come to mind, not morphology. Of course, from a sociobiological point of view, behavior could be considered to be just another aspect of an organism's phenotype, the result of the interplay between the organism's genetic legacy and the environment the organism finds itself in, and "phenotype" is "morphology."

One area of research in which behavior indisputably results in "morphology" is the study of trace fossils. Richter (1924) first proposed an optimization model to explain the geometry of the various spirals, meanders, and loops found on bedding planes in the fossil record, which are the record of the activity of, in many cases, unknown organisms. Richter assumed that many of these trace fossils were produced by organisms as they fed and, in accordance with natural selection theory, proposed that the observed geometries resulted from organisms seeking to maximize their coverage of a given area of seafloor while minimizing the length of the pathway they had to take to optimize their food uptake. In addition, Richter proposed that feeding behavior in two dimensions, such as on the seafloor, could be described in terms of three parameters or basic reactions: (1) "strophotaxis," a reaction in which the organism makes a turn of around 180° and heads back in the general direction it previously was heading away from, (2) "phobotaxis," a reaction in which the organism avoids crossing the tracks made by either other organisms or itself, and (3) "thigmotaxis," a reaction in which the organism seeks to remain close to its previous trackway.

Computer simulation of feeding behavior came nearly a half-century later in the theoretical morphologic study of Raup and Seilacher (1969). They made Richter's behavioral three-reaction model (1924) much more flexible and lifelike by adding six additional "behavioral characteristics": (1) the turning radius for strophotactic turns, (2) the strength of the thigmotactic reaction, measured as the mean distance between old and new tracks, (3) the range of deviation permitted for the strength of the thigmotactic reaction, (4) the relative strength of thigmotactic and phobotactic reactions, measured as the angle made in a turn toward an old track relative to the distance of the new track away from the old track, (5) the mean length of meander segments, and (6) the range of deviation permitted for the length of meander segments. Raup and Seilacher were successful in simulating what appear to be exceedingly complex labyrinthine pathway-geometries with this model, pathway geometries that appeared similar to those found in the fossil record and also that apparently had not been produced by organisms in nature. Thus they achieved the theoretical morphologic goal of simulating both existent and nonexistent morphology, although the "morphology" in this case was behavior.

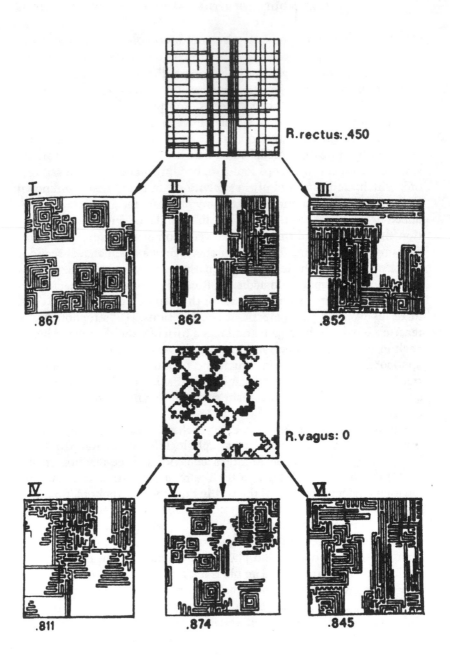

R.rectus: .450

I.

.867

II.

.862

III.

.852

R.vagus: 0

IV.

.811

V.

.874

VI.

.845

Papentin (1973a,b) and Papentin and Röder (1975) went further and sought to simulate the evolution of behavior. They designed a modified version of the Richter (1924) and Raup and Seilacher (1969) models of behavior and then added intergenerational time by allowing the parameters to "evolve" by means of mutation and/or recombination. The process of natural selection was added to the evolution simulations by giving a greater probability of "reproduction" to those organisms that produced trackways that less frequently crossed previously existing trackways. They found that all their hypothetical organisms, regardless of the type of trackway with which the organism was initially programmed, evolved both turning behavior (a form of strophotaxis) and trackway contact-maintenance behavior (thigmotaxis). Beyond that evolutionary threshold, the simulations were more variable, with some organisms evolving spiral trackways, others evolving meander trackways, and with still others evolving a combination of both (figure 10.4). The conclusion of their theoretical morphologic analysis of behavior was that the avoidance of pathway crossing alone (phobotaxis) was sufficient to account for the majority of behaviors seen in two-dimensional feeding trackways.

Other theoretical models of feeding behavior are briefly reviewed in Telford (1990), but I do not include them here, as they do not, in general, involve "morphology." The deposit-feeding model that Telford developed could itself be considered "theoretical morphologic," as it depends largely on the morphology of podia in sand dollars and on the morphology (grain size) of substrates encountered by sand dollars during feeding. More important, however, Telford's study was conducted in the spirit of a theoretical morphologic analysis, as the purpose of the model simulation was to allow "computer prediction of the spectrum of ingested grain sizes from any given substrate area, real or

FIGURE 10.4. Computer simulation of the evolution of behavior in two hypothetical benthic worm species, *Rectangulus rectus* (upper half of the figure) and *Rectangulus vagus* (lower half of the figure). In the initial configuration, *R. rectus* is capable only of moving straight ahead, whereas *R. vagus* turns randomly after each step in the simulation. The process of natural selection was then simulated by giving a greater probability of "reproduction" to those organisms that produced trackways that crossed less frequently. Numbers under the simulations indicate the effectiveness of the hypothetical worms in substrate coverage with minimum path crossing in the last 60 of 180 generations. *After F. Papentin and H. Röder (1975), Neues Jahrbuch für Geologie und Paläontologie, Abhandlungen, 148 and used with permission.*

imaginary" (Telford 1990:77), with the ultimate goal of allowing "a theoretical comparison of feeding behavior between species which do not normally co-occur" (Telford 1990:75). And such a goal, the creation and comparison of a spectrum of both existent and nonexistent form, is pure theoretical morphology.

# XI

# The Future of Theoretical
# Morphology

This chapter will examine some historical trends during the past 30 years with the aim of understanding why the discipline of theoretical morphology has developed in the way that it has. First I shall look at the two major foci of theoretical morphology, theoretical morphogenetic modeling and theoretical morphospace analyses. At the end of the chapter I shall briefly summarize where the field of theoretical morphology has been and point out the directions I consider most promising for the field to go in the future.

The future of theoretical morphology lies with the graduate students of the present and future generations. With that fact in mind, I have attempted in this book to point out many potential Ph.D. dissertation topics in theoretical morphology, intellectually challenging problems that await the attention of young graduate student researchers. So, if you are a graduate student considering a career in morphology or a professional morphologist with graduate students hunting for research topics, keep your pencils at the ready!

## THE PURSUIT OF THEORETICAL
## MORPHOGENETIC MODELING

Two challenges for theoretical morphogenetic modelers appear to be clear, at least to me. The teaching of morphogenetic modeling needs to be more actively promoted, and the practitioners of theoretical morphogenetic modeling need to develop closer ties with experimentalists.

Theoretical morphogenetic modeling has suffered these past 30 years from a form of pedagogic neglect. There are many fine morphometrics textbooks, and most graduate university curricula in biology and paleobiology offer courses in morphometrics. In addition, the variety of morphometric software is abundant and is easily available. The same cannot be said of theoretical morphology, however. I began the preface of this book with a quotation from Enrico Savazzi, who lamented the lack of instruction in theoretical morphology in modern university studies. Consider another quotation, and a proposal for the future, from Savazzi (1995:238, italics mine):

> Like all computer-based fields, theoretical shell morphology requires hands-on experience to be appreciated. However, very little computer software for this field is publicly available. . . . Since the development of a field of research is dependent on a sufficient number of scientists becoming interested in it, I am proposing the notion that *the development of software for specific use as a teaching aid in theoretical shell morphology is as essential as leading-edge research* to ensure the continued development of this field.

I entirely agree with Savazzi that theoretical morphogenetic modeling is best appreciated as a hands-on experience, and in the example in chapter 8, I tried to show how much fun it can be. Interactive software, rather than the passive equations given in chapter 8, would undoubtedly make the learning experience more exciting. It is true that very few theoretical morphologists have actually published their mathematical models in the form of computer programs that others might readily use, modify, and experiment with. The published programs of Savazzi (1985, 1990b, 1993) and Okamoto (1988a) are notable and commendable exceptions, as is making the software available through an established publishing firm or professional society, which Savazzi (1990c) did. Making software available "by writing to the author of the paper" (e.g., Ackerly 1989a, Cortie 1989, Stone 1995) is undesirable, as the professional addresses of individual researchers (and likewise the Internet) tend to be much less permanent than those of firms or societies.

The published programs and software that are available for theoretical morphogenetic modeling are not designed specifically for teaching but, rather, for advanced simulation studies. Accordingly, Savazzi (1995) challenged theoretical morphologists to develop software specifically for teaching theoretical morphogenetic modeling.

A second point of view concerns the lack of formal teaching of the techniques of theoretical morphology at most universities, in addition to the lack of teaching materials. Has theoretical morphogenetic modeling in the past 30 years failed to produce fruitful results? Consider Hickman's assessment (1993b:171), who considers theoretical morphogenetic modeling's original objective to have been simulation:

> Theoretical morphology abandoned this objective rather early in its history. If I may hazard a guess as to the reason, it is that simulation modeling, the original tool of theoretical morphology (Raup 1961, 1962), is not a fruitful means of discovering simple and general principles of structural diversity. Theoreticians thrive on creating more complex models.

One of the greatest surprises of early theoretical morphologic modeling was that very complex organic forms could be produced by relatively simple mathematical models. From this initial result, however, it is true that theoreticians have tended to create ever more complex models in an attempt to more closely simulate natural growth systems (this can be clearly seen for models of accretionary growth systems in both chapters 8 and 9). Yet is this not the fundamental goal of a model of morphogenesis, to simulate the growth system under study? In contrast to Hickman's assessment of simulation modeling, Harrison (1987) bemoans the scarcity of theoretical models of morphogenesis in developmental biology, stating that to pursue a career in developmental biology "is akin to doing quantum physics before Schrödinger and Heisenberg, or genetics before Mendel" (Harrison 1987:370).

A much more serious problem with theoretical models of morphogenesis may be their apparent lack of definitive predictive power. This ambiguity can be seen most clearly in chapter 10, in which both the reaction-diffusion models and the neurokinetic models of the morphogenesis of shell ornamentation produce similar results, even though the models' fundamental biological assumptions are very different. To make matters worse, the same shell pigmentation patterns can be produced by cellular automata, which are vastly different in mathematical concept from the continuous differential equations used in physicochemical developmental models.

To address this problem, theoretical morphogenetic modelers may need to work more closely with experimentalists. Jacobson (1980:669) argued that the potential benefits of modeling to experimentalism (in this case, embryology) were as follows:

> The development of a computer simulated model requires that assumptions and steps in reasoning be stated explicitly. This process brings added rigor to analysis of the biological system, improved observations of the embryo itself, and suggestions for new experiments. An interplay develops between the two systems, the actual and the simulated, in which hypotheses and experimental results from each system can be used to modify the perceptions of the other. A successful simulation can be used to simulate experiments that may not be possible on the embryo.

Harrison (1987:383) echoed this sentiment:"The heart of the scientific enterprise, especially for fields lacking consensus on the kind of theory needed to explain the data, is close interaction between experiment and theory. In morphogenesis, this has been lamentably inadequate, but it seems to be beginning."

## THE PURSUIT OF THEORETICAL MORPHOSPACE ANALYSES

In contrast to the mixed reviews of theoretical morphogenetic modeling, theoretical morphospace analyses have drawn much more attention and scientific acclaim. Raup's (1966, 1967) initial theoretical morphospace analyses certainly made quite an intellectual splash, and most textbooks on morphology either figure or discuss the famous "Cube" $W-D-T$ theoretical morphospace of univalved shell form. The discipline has advanced as other morphologists have constructed other theoretical morphospaces over the past 30 years, and the fruits of their labors are the subjects of chapters 3 through 6. Yet still the number of morphologists who have conducted theoretical morphologic analyses is much smaller than the number of morphologists who have chosen to embrace empirical morphospaces rather than theoretical morphospaces. Why?

One answer that comes to mind is methodological difficulty. Theoretical morphospaces are often not easy to construct or to construct without flaws (such as the lack of algebraic independence of the dimensions of Raup's own *W–D–T* theoretical morphospace, as pointed out in chapter 4). But empirical morphospace construction is simplicity itself. The worker need only take lots and lots of morphologic measurements on a series of organisms, dump those measurements into a computer, specify a particular multidimensional ordination program, and the computer prints out an impressive hyperdimensional empirical morphospace. Then the morphologist has the happy task of writing a research paper explaining why certain organisms cluster over here in the space as opposed to over there with those other organisms, and so on.

Have microcomputers made empirical morphospace analyses too easy? I can remember, back in my graduate student days, when microcomputers were essentially toys. They were fun to play with, but you could not perform serious complex analyses with them. To do empirical morphospace analyses in those days required using a mainframe computer, which often was woefully overworked with everyone else's computations, which meant that you often had to wait a day or two for your computer printout of results. And often you found that you had made a mistake in the programming and that the entire job had to be resubmitted with another one- to two-day turnaround delay, not to mention the hours and hours spent in the initial card punching of both the data and programming, when a single typographical mistake on a punched card meant the entire program would crash. I would predict that many readers do not even know what a card-punching machine is. When I was conducting my doctoral research, Dave Raup himself was using a black-and-white television set as a computer monitor, linked by a modem to the mainframe computer. We (the graduate students) all thought it was great, as it freed us from the infernal card-punching machines. And if the mainframe computer was down (as it often was), you could simply turn the television to another channel and watch the news.

The advent of high-powered and inexpensive microcomputers has changed all that. Morphologists today can perform extremely complicated multidimensional ordination analyses in their own lab with their own machine. Turnaround times of days have been reduced to minutes, and actual required computational times of minutes have been reduced to nanoseconds. And with peripheral devices such as digitizers and scanners, even the morphologic measurements themselves can be virtually performed by the computer.

281

The use of computers to generate hypothetical, biologically nonexistent morphology was one of the most impressive early accomplishments of theoretical morphology. It is thus ironic that the computer—specifically the microcomputer—may actually have hampered the potential development of the field of theoretical morphology. It is undeniable that theoretical morphospaces simply do not exist for many types of organisms and growth systems. Yet rather than meeting this challenge head on and working (and thinking) to construct new theoretical morphospaces to analyze these organisms and growth systems, many morphologists simply choose to take the less demanding route and conduct an empirical morphospace analysis of the organisms instead.

The nature and shape of the hyperspace in a theoretical morphospace are determined by geometry, not by organisms. The dimensions of a theoretical morphospace exist mathematically pure and pristine in the absence of any measurement data. Such a morphospace can not only reveal what *could be* in nature (regardless of what nature has actually produced), it can also reveal *what cannot exist* at all. Such a morphospace may be a bit difficult to construct, but the benefits surely outweigh the cost of the effort.

## THERE IS SO MUCH TO BE DONE! (OR HINTS FOR PROMISING YOUNG GRADUATE STUDENTS)

Table 11.1 summarizes the groups of organisms that have been the subject of theoretical morphospace analyses and that were discussed in chapters 3 through 7 (see also Reif and Weishampel 1991 for an extensive list of theoretical morphologic, as well as morphometric, analyses conducted over the past 30 years). As table 11.1 shows, fully two-thirds of the *Baupläne* of life on Earth have never been examined using the techniques of theoretical morphology. Of the one-third that have, only a few (such as the Mollusca) have undergone analyses of nearly the total spectrum of form exhibited by organisms of that *Bauplan*. The majority of *Baupläne* have been examined only in part (such as the Chordata), leaving a vast range of morphology in those *Baupläne* yet to be looked at rigorously.

### Needed: Students to Explore Theoretical Morphospaces

In chapters 3 through 6, we examined the theoretical morphospaces that have been extensively explored, such those created for ectocochleate cephalopods, gastropods, and brachiopods. We also looked at the-

**TABLE 11.1.** The major structural types, or *Baupläne*, of life on Earth and a list of those that have been the subject, at least in part, of theoretical morphospace analyses. *Classification of life modified from Whittaker 1969 and Levin 1996.*

| Bauplan | Theoretical Morphospace Analysis |
|---|---|
| Kingdom Monera | |
| Phylum Methanocretrices | none |
| Phylum Omnibacteria | none |
| Phylum Cyanobacteria | Chapter 6: stromatolites |
| Phylum Myxobacteria | none |
| Phylum Aphragmabacteria | none |
| Phylum Spirochaetae | none |
| Phylum Actinobacteria | none |
| Kingdom Protoctista | |
| Phylum Euglenophyta | none |
| Phylum Xanthophyta | none |
| Phylum Chrysophyta | Chapter 6: silicoflagellates |
| Phylum Pyrrophyta | none |
| Phylum Hyphochytridiomycota | none |
| Phylum Plasmodiophoromycota | none |
| Phylum Sporozoa | none |
| Phylum Cnidosporidia | none |
| Phylum Zoomastigina | none |
| Phylum Sarcodina | Chapter 6: foraminiferids |
| Phylum Ciliophora | none |
| Kingdom Plantae | |
| Division Rhodophyta | none |
| Division Phaeophyta | none |
| Division Chlorophyta | Chapter 6: receptaculitids |
| Division Charophyta | none |
| Division Bryophyta | none |
| Division Psilophyta | Chapter 3: pteridophytes |
| | Chapter 7: pteridophytes |
| Division Lycopodophyta | none |
| Division Equisetophyta | none |
| Division Polypodiophyta | none |
| Division Pinophyta | Chapter 3: gymnosperms |
| | Chapter 7: gymnosperms |
| Division Magnoliophyta | Chapter 3: angiosperms |
| | Chapter 7: angiosperms |
| Kingdom Fungi | |
| Division Myxomycophyta | none |
| Division Eumycophyta | none |

TABLE 11.1. Continued

| Bauplan | Theoretical Morphospace Analysis |
|---|---|
| Kingdom Animalia | |
| Phylum Porifera | Chapter 6: stromatoporoids |
| Phylum Archaeocyatha | none |
| Phylum Cnidaria | Chapter 6: corals |
| Phylum Ctenophora | none |
| Phylum Platyhelminthes | none |
| Phylum Nemertea | none |
| Phylum Nematoda | none |
| Phylum Acanthocephala | none |
| Phylum Nematomorpha | none |
| Phylum Rotifera | none |
| Phylum Gastrotricha | none |
| Phylum Bryozoa | Chapter 3: fenestrates, cheilostomes |
| Phylum Brachiopoda | Chapter 4: articulates |
| | Chapter 7: articulates |
| Phylum Phoronida | none |
| Phylum Annelida | none |
| Phylum Onychophora | none |
| Phylum Arthropoda | none |
| Phylum Mollusca | Chapters 4: cephalopods, gastropods |
| | Chapter 5: bivalves |
| | Chapter 7: cephalopods |
| Phylum Hyolitha | none |
| Phylum Pogonophora | none |
| Phylum Chaetognatha | none |
| Phylum Echinodermata | Chapter 6: echinoids |
| Phylum Hemichordata | Chapter 3: graptolites |
| | Chapter 6: graptolites |
| Phylum Conodonta | none |
| Phylum Chordata | Chapter 6: chondrichthyans |
| | Chapter 7: chondrichthyans |

oretical morphospaces that have yet just barely been entered and that clearly are awaiting further exploration, such as those for bryozoan colonies (McKinney and Raup 1982, Cheetham and Hayek 1983, Starcher 1987) and tropical trees (Honda and Fisher 1978).

We also found theoretical morphospaces that have been created but never explored. For example, we examined perfectly fine theoretical morphospaces of bivalve mollusc form (Savazzi 1987), land plant form (Niklas and Kerchner 1984), foraminiferal form (Berger 1969, Signes et al. 1993), silicoflagellate form (McCartney and Loper 1989), and shark dermal-skeleton form (Reif 1980). In all cases, the creators of these morphospaces indicated, to a greater or lesser degree, where they expect major taxonomic groupings of bivalve molluscs, land plants, foraminifers, silicoflagellates, and sharks to be in those theoretical morphospaces. That is, they each gave a rough indication of the regions of morphospace that are believed to be occupied by all those organisms. Thus, for all these existing theoretical morphospaces, we are in the same position as we were back in 1966 when Raup created the "Cube" morphospace of univalved shell form (figure 4.3). We have a glimmering of an impression of what nature has produced, but we do not really know for certain.

The actual *exploration* of these theoretical morphospaces has yet to be undertaken: a series of excellent graduate dissertation topics are simply waiting to be seized by students of evolutionary morphology. As a first consideration, the authors of these theoretical morphospaces may be mistaken in their assessments of the regions of morphospace occupied by various organisms. For example, as chapter 4 pointed out, Raup (1966) vastly underestimated the distribution of gastropod form in the $T$-dimension of the $W–D–T$ theoretical morphospace (figure 4.3).

Second, we have absolutely no idea of the density of distribution of organic forms in these theoretical morphospaces. Even if each author's expected distributions of organic form in these theoretical morphospaces turn out to be more or less the real distributions in nature, we have no idea where the morphologic density "peaks" and "valleys" are. And without knowing where those density clusters of form are in the theoretical morphospaces, we have no way to analyze their functional significance.

Third, the dimensionality of these theoretical morphospaces has not been put to the test of practicality. As we saw in chapter 4, the dimensions of the $W–D–T$ theoretical morphospace work well for the analysis of the distribution of real morphology in the ammonoids but are both awkward and somewhat uninformative in the analysis of the distribution of real morphology in the gastropods. Accordingly, gastropod workers have been motivated to modify Raup's (1966) theoretical morphospace in order to add dimensions whose parameter values are easier to measure on real organisms or to add dimensions for aspects of morphology important to gastropods (such as aperture orientations) but not included in Raup's original parameterization.

Last, we obviously cannot examine the addition of the time dimension to these theoretical morphospaces (as we did in chapter 7), to observe the pattern of the evolution of organic form in theoretical morphospace through time, without having first mapped the distribution of existent and nonexistent form in these theoretical morphospaces.

### Needed: Students to Create Theoretical Morphospaces

Creating a theoretical morphospace is usually more difficult than exploring a theoretical morphospace, but not necessarily much more difficult. In many cases, it simply requires a little thought. The absence of a theoretical morphospace for a particular *Bauplan* (table 11.1) may also simply mean that no one has bothered to think about creating such a theoretical morphospace yet. It would not take much effort, for example, to create a theoretical morphospace that would contain hypothetical "worms." Worms are, after all, only very long cylinders (to a first approximation). Quite a few animal phyla (table 11.1) could be examined in such a simple "worm space." If one added a little complexity, such as segmenting the cylinders (a long cylinder composed of many smaller cylinder subunits), one would be well on the way to creating a theoretical morphospace in which the evolution of the Annelida could be examined. And from such an annelid space, one could begin, perhaps, to attack a major animal group whose complex morphology has thus far defied theoretical morphospace construction: the Arthropoda.

We also have encountered groups for which no theoretical morphospace currently exists but for which do exist theoretical models of morphogenesis whose parameterization might be adapted to the construction of such a theoretical morphospace, or groups for whom a specific model of morphogenesis does not exist but whose growth and geometry is similar to groups for whom models do exist. Examples include the colonial cnidarians, the radiolarians, the poriferans, and the echinoderms.

## A THEORETICAL MORPHOLOGIC CHALLENGE

*I believe that the question of defining morphospaces and mapping their differential filling through time is so vital to our understanding of life's history, particularly to the potential contribution of paleontologists. Yet relatively little has been done in this area.*

Gould (1991:422)

Just how little has been done in the area of theoretical morphospace analyses can be seen by examining table 11.1. I urge my colleagues and fellow students of evolutionary morphology to consider table 11.1

as a list of potential Ph.D. dissertations. There are *Baupläne* in table 11.1 whose total range of morphology could easily be examined using existing theoretical morphospaces or modifications of existing theoretical morphospaces. There are *Baupläne* in table 11.1 for which theoretical morphospaces could easily be constructed using theoretical morphogenetic models that already exist, using parameterizations that have already been formulated. And last, there are *Baupläne* in table 11.1 that thus far have defied theoretical morphospace analysis (such as the Arthropoda). The intellectual challenge is clear—will not someone accept it?

*Adaptation:* (1) The state of being suited or fitted. (2) The process of adjusting to fit new conditions. (3) In biology, an advantageous conformation of an organism to changes in its environment. (Modified from Funk & Wagnalls Standard College Dictionary).

*Allometric:* In studies of ontogeny, denotes a particular type of anisometric growth (q.v.) in which the relative growth of two different aspects of form in an organism (e.g., $X$ and $Y$) follow a power function of the type: $Y = bX^a$, where $a$ and $b$ are constants and the value of $a$ is not equal to 1.

*Analogous:* Morphologic characters in different organisms that may be similar in structure and function but that have different evolutionary origins (cf. Homologous).

*Anisometric:* In studies of ontogeny, denotes growth processes in which an organism's form or proportions change with time (cf. Isometric).

*Apomorph:* A "derived character" in cladistic analyses, a morphological character that is a new evolutionary novelty in a lineage. If possessed by a single species only (or defining a single branch in a cladogram), such characters are termed "autapomorphies" (cf. Synapomorph, Plesiomorph).

*Bauplan* (plural: *Baupläne*): A German engineering term translated simply as "blueprint." In biology, the term is used to represent the basic architectural and organizational pattern shared by the members of a monophyletic clade of higher taxonomic rank. According to Valentine (1986:209), "At the upper levels of the taxonomic hierarchy, phyla- or class-level clades are characterized by their possession of particular assemblages of homologous architectural and structural features; in this paper, it is to such assemblages that the term Bauplan is applied."

*Biodisparity:* The degree of difference, unlikeness, or distinctiveness seen in the range of morphologies or anatomical designs present in a clade (cf. Biodiversity, Morphologic diversity).

*Biodiversity:* The number of taxa (species, genera, and so on) that are found in a given clade, geographic area, or ecological unit (community, province,

**288**

and so on). Often used as a synonym for taxonomic diversity (cf. Biodisparity, Morphologic diversity).

*Clade:* A monophyletic group of taxa, that is, a group of species (genera, families, and so on), whose members all have a common ancestor.

*Constructional morphology:* A school of morphologic analysis founded by Adolf Seilacher in 1970. In constructional morphologic analyses, organic form is postulated to result from the interplay of three factors: functional (adaptational) constraints, fabricational (morphogenetic) constraints, and historical (phylogenetic) constraints (cf. Functional morphology, Theoretical morphology).

*Convergent evolution:* The evolution of morphologies in different organisms that are very similar in structure and function, even though those organisms do not descend from a common ancestor (cf. Iterative evolution).

*Empirical morphospaces:* "Multidimensional morphological spaces produced from the mathematical analysis of actual measurement data using such techniques as principal components analysis, factor analysis, Fourier analysis, or other polynomial series approximations of natural morphology" (McGhee 1991:96). Empirical morphospaces are explicitly produced with, and are dependent on, actual measurement data from existent organic form. Such morphospaces have no existence in the absence of actual measurement data (cf. Theoretical morphospaces).

*Evolution:* Any change in the gene frequencies in a population with time (modified from Wilson and Bossert 1971).

*Functional morphology:* A school of morphologic analysis that specifies the functional or adaptive aspects of organic form, usually by employing the paradigm methodology (q.v.) of M. J. S. Rudwick (cf. Constructional morphology, Theoretical morphology).

*Genotype:* The genetic makeup of an individual organism, as coded for in its DNA (cf. Phenotype).

*Gnomonic growth:* Growth in which each new growth increment or unit is a gnomon to previous growth increments or units; that is, growth takes place without any change in shape. A synonym for isometric growth (q.v.).

*Holophyletic:* A monophyletic taxonomic grouping in which all descendants of a single common ancestor are included in the group (cf. Paraphyletic).

*Homologous:* Morphologic characters in different organisms that may differ in structure or function but that nevertheless have a common evolutionary origin (cf. Analogous).

*"Hybrid morphospace":* A morphospace containing both empirical input data, taken from actual specimens, and theoretical model parameters, which are used to manipulate the input data in a theoretical fashion to produce a hypothetical spectrum of form. Such a morphospace is technically neither empirical nor theoretical but a mix of both (cf. Empirical morphospaces, Theoretical morphospaces).

*Hyperspace:* A conceptual space, as opposed to experiential space, whose dimensionality usually exceeds three dimensions. Hyperspaces are usually defined mathematically, but this is not a formal requirement.

*Isometric:* In studies of ontogeny, denotes growth processes in which an organism's form and proportions do not change with time (cf. Anisometric).

*Iterative evolution:* Repeated evolution of different descendant taxa, all with very similar morphologies, from a single common ancestral taxon (cf. Convergent evolution).

*Ma: Mega annum* (plural: *Mega annos*), 1 million years ($10^6$ yr). May refer both to a duration of time or a date before the present; the latter is formally designated Ma BP.

*Monophyletic:* A natural taxonomic grouping whose members all have a single common ancestor (cf. Polyphyletic).

*Morphologic diversity:* The number of different morphologies found in a given clade, geographic area, or ecological unit (cf. Biodisparity, Biodiversity).

*Morphometrics:* A school of morphologic analysis concerned with the precise measurement of form in individual organisms and with the precise comparison of those measurements among different individuals.

*Natural selection:* The differential change in genotypic frequencies in a population with time due to the differential reproductive success of their phenotypes (modified from Wilson and Bossert 1971).

*Ontogeny:* Description of the growth and developmental changes that take place in the life of a single individual organism (cf. Phylogeny).

*Paradigm methodology:* A formalized technique of mechanical analogy in investigating the functional significance of an organic structure or form, designed by Martin J. S. Rudwick in 1964. Several functions are first postulated for the structure under investigation, and then the optimal mechanical design for each function is determined (the paradigm). The paradigm form that most closely matches the original structure is considered to demonstrate the actual function of the structure.

*Paraphyletic:* A monophyletic taxonomic grouping in which some of the descendants, but not all the descendants, of a single common ancestor are included in the group (cf. Holophyletic).

*Phenotype:* The characteristics of an individual organism produced by the interaction between its genotype and its environment during the span of its ontogeny. Phenotype includes not only the physical morphology of an organism (the subject of this book) but also its behavior (cf. Genotype).

*Phylogeny:* Description of the evolutionary changes that take place in a monophyletic group of organisms in geologic time (cf. Ontogeny).

*Plesiomorph:* A "primitive character" in cladistic analyses, a morphological character that is not a new evolutionary feature but has simply been inherited from an ancestor (cf. Apomorph).

*Polyphyletic:* An unnatural taxonomic grouping whose members are descended from different ancestors (cf. Monophyletic).

*Species:* (1) A grouping of individual organisms that share the same gene pool (i.e., that can freely interbreed with one another and produce fertile offspring). The species is the fundamental unit of the taxonomic hierarchy. (2) The lowest level in the classification of life. All organisms are given a two-part scientific name consisting of its genus and species designations.

*Symplesiomorph:* A "shared primitive character" in cladistic analyses, an ancient morphological character present in several species or clades of species simply due to its inheritance from a remote ancestor (cf. Synapomorph).

*Synapomorph:* A "shared derived character" in cladistic analyses, a recent morphological character present in several species due to its inheritance from the species' latest common ancestor. Synapomorphies are homologous characters (q.v.), and the species possessing them are termed a "sister-group" (cf. Symplesiomorph).

*Taxon (plural: taxa):* A general term designating all organisms at a given level in the classification of life. The main Linnaean classification levels, from most specific to most general, are species, genus, family, order, class, phylum, and kingdom.

*Taxonomic diversity:* The number of taxa (species, genera, and so on) found in a given clade, geographic area, or ecological unit (community, province, and so on). Often used as a synonym for biodiversity (cf. Biodisparity, Morphologic diversity).

*Theoretical design spaces:* Theoretical hyperspaces that include dimensions that are not strictly morphological. A theoretical design space "differs from a morphospace by combining morphological axes with axes that are ecological, behavioral, or physiological" (Hickman 1993b:170) (cf. Empirical morphospaces, Theoretical morphospaces).

*Theoretical morphology:* A school of morphologic analysis that was founded by David M. Raup and that is the subject of this book. In theoretical morphologic analyses, organic form is simulated by using geometric or other mathematical models, producing either theoretical morphospaces (hypothetical form distribution) or theoretical morphogenesis (hypothetical growth models) (cf. Constructional morphology, Functional morphology).

*Theoretical morphospaces:* "N-dimensional geometric hyperspaces produced by systematically varying the parameter values of a geometric model of form" (McGhee 1991:87). Theoretical morphospaces are produced without any measurement data from real organic form. They not only are independent of existent morphology, but they also can be used to produce nonexistent morphology. Such morphospaces exist in the absence of any measurement data whatsoever (cf. Empirical morphospaces).

# REFERENCES

Abelson, H., and A. A. DiSessa. 1982. *Turtle Geometry.* Cambridge, Mass.: MIT Press.

Ackerly, S. C. 1989a. Kinematics of accretionary shell growth, with examples from brachiopods and molluscs. *Paleobiology* 15:147–164.

Ackerly, S. C. 1989b. Shell coiling in gastropods: Analysis by stereographic projection. *Palaios* 4:374–378.

Ackerly, S. C. 1990. Using growth functions to identify homologous landmarks on mollusc shells. In F. J. Rohlf and F. L. Bookstein, eds., *Proceedings of the Michigan Morphometrics Workshop.* Special Publications no. 2:339–344. Ann Arbor: University of Michigan Museum of Zoology.

Ackerly, S. C. 1992a. Morphogenetic regulation in the shells of bivalved organisms: Evidence from the geometry of the spiral. *Lethaia* 25:249–256.

Ackerly, S. C. 1992b. The structure of ontogenetic variation in the shell of *Pecten. Palaeontology* 35:847–867.

Bard, J. 1990. The fifth day of creation. *Bioessays* 12:303–306.

Bard, J. B. L. 1981. A model for generating aspects of zebra and other mammalian coat patterns. *Journal of Theoretical Biology* 93:363–385.

Batschelet, E. 1974. *Introduction to Mathematics for Life Scientists.* Berlin: Springer-Verlag.

Bayer, U. 1977. Cephalopoden-Septen Teil 2: Regelmechanismen im Gehäuse- und Septenbau der Ammoniten. *Neues Jahrbuch für Geologie und Paläontologie, Abhandlungen* 155:162–215.

Bayer, U. 1978a. A growth model for Foraminifera. In A. Seilacher and R. Westphal, eds., *Report 1976–1978: Sonderforschungsbereich 53 Tübingen. Neues Jahrbuch für Geologie und Paläontologie, Abhandlungen* 157:65–69.

Bayer, U. 1978b. Morphogenetic programs, instabilities, and evolution—A theoretical study. *Neues Jahrbuch für Geologie und Paläontologie, Abhandlungen* 156:226–261.

Bayer, U. 1985. *Pattern Recognition Problems in Geology and Paleontology.* Berlin: Springer-Verlag.

Bayer, U., and G. R. McGhee Jr. 1984. Iterative evolution of Middle Jurassic ammonite fauna. *Lethaia* 17:1–16.

Bayer, U., and G. R. McGhee Jr. 1985a. Evolution in marginal epicontinental basins: The role of phylogenetic and ecological factors (ammonite replacements in the German Lower and Middle Jurassic). In U. Bayer and A. Seilacher, eds., *Sedimentary and Evolutionary Cycles*, Lecture Notes in Earth Sciences 1:164–220. Berlin: Springer-Verlag.

Bayer, U., and G. R. McGhee Jr. 1985b. Evolution of Middle Jurassic ammonites: A reply. *Lethaia* 18:38.

Bell, A. D. 1986. The simulation of branching patterns in modular organisms. *Philosophical Transactions of the Royal Society of London, Series B* 313: 143–159.

Bell, A. D., D. Roberts, and A. Smith 1979. Branching patterns: The simulation of plant architecture. *Journal of Theoretical Biology* 81:351–375.

Berger, W. H. 1969. Planktonic foraminifera: Basic morphology and ecologic implications. *Journal of Paleontology* 43:1369–1383.

Bookstein, F. L. 1977a. Orthogenesis of the hominids: An exploration using biorthogonal grids. *Science* 197:901–904.

Bookstein, F. L. 1977b. The study of shape transformation after D'Arcy Thompson. *Mathematical Biosciences* 34:177–219.

Brasier, M. D. 1980. *Microfossils*. London: Allen & Unwin.

Brasier, M. D. 1982a. Architecture and evolution of the foraminiferid test—A theoretical approach. In F. T. Banner and A. R. Lord, eds., *Aspects of Micropalaeontology*, pp. 1–41. London: Allen & Unwin.

Brasier, M. D. 1982b. Foraminiferid architectural history; a review using the MinLOC and PI methods. *Journal of Micropaleontology* 1:95–105.

Brasier, M. D. 1984. Some geometrical aspects of fusiform planispiral shape in larger foraminifera. *Journal of Micropaleontology* 3:11–15.

Brasier, M. D. 1986. Form, function, and evolution in benthic and planktic foraminiferid test architecture. In B. S. C. Leadbeater and R. Riding, eds., *Biomineralization in Lower Plants and Animals, the Systematics Association Special vol. no.* 30:251–268. Oxford: Clarendon Press.

Briggs, D. E. G., and R. A. Fortey. 1989. The early radiation and relationships of the major arthropod groups. *Science* 246:241–243.

Briggs, D. E. G., R. A. Fortey, and M. A. Wills. 1992. Morphological disparity in the Cambrian. *Science* 256:1670–1673.

Burnaby, T. P. 1966. Allometric growth of ammonoid shells: A generalization of the logarithmic spiral. *Nature* 209:904–906.

Cain, A. J. 1977. Variation in the spire index of some coiled gastropod shells, and its evolutionary significance. *Philosophical Transactions of the Royal Society of London (B: biological sciences)* 277:377–428.

Cain, A. J. 1978a. The deployment of operculate land snails in relation to shape and size of shell. *Malacologia* 17:207–221.

Cain, A. J. 1978b. Variation of terrestrial gastropods in the Philippines in relation to shell shape and size. *Journal of Conchology* 29:239–245.

Cain, A. J. 1980. Whorl number, shape, and size of shell in some pulmonate faunas. *Journal of Conchology* 30:209–221.

293

Chamberlain, J. A., Jr. 1976. Flow patterns and drag coefficients of cephalopod shells. *Palaeontology* 19:539–563.

Chamberlain, J. A., Jr. 1980. Hydromechanical design of fossil cephalopods. In M. R. House and J. R. Senior, *The Ammonoidea, Systematics Association Special Volume* 18:289–336.

Chamberlain, J. A., Jr., and J. S. Weaver. 1989. Cephalopod theoretical morphology: A new look at an old idea. *Geological Society of America, Abstracts with Programs* 21(6):289–290.

Chapman, R. E., D. Rasskin-Gutman, and D. B. Weishampel. 1996. Exploring the evolutionary history of a group using multiple morphospaces of varying complexity and philosophy. In J. E. Repetski, ed., *Sixth North American Paleontological Convention Abstracts of Papers*. Paleontological Society Special Publication no. 8:66. Washington, D.C.: Smithsonian Institution and the Paleontological Society.

Checa, A. 1987. Morphogenesis in ammonites—Differences linked to growth pattern. *Lethaia* 20:141–148.

Checa, A. 1991. Sectorial expansion and shell morphogenesis in molluscs. *Lethaia* 24:97–114.

Checa, A., and R. Aguado. 1992. Sectorial-expansion analysis of irregularly coiled shells: Application to the Recent gastropod *Distorsio*. *Palaeontology* 35:913–925.

Cheetham, A. H., and L. C. Hayek. 1983. Geometric consequences of branching growth in adeoniform Bryozoa. *Paleobiology* 9:240–260.

Cheetham, A. H., L. C. Hayek, and E. Thomsen. 1980. Branching structure in arborescent animals: Models of relative growth. *Journal of Theoretical Biology* 85:335–369.

Cheetham, A. H., L. C. Hayek, and E. Thomsen. 1981. Growth models in fossil arborescent cheilostome bryozoans. *Paleobiology* 7:68–86.

Cheetham, A. H., and E. Thomsen. 1981. Functional morphology of arborescent animals: Strength and design of cheilostome bryozoan skeletons. *Paleobiology* 7:355–383.

Cohen, D. 1967. Computer simulation of biological pattern generation processes. *Nature* 216:246–248.

Cortie, M. B. 1989. Models for mollusc shape. *South African Journal of Science* 85:454–460.

Cortie, M. B. 1993. Digital seashells. *Computers and Graphics* 17:79–84.

Cowen, R., and M. J. S. Rudwick. 1967. *Bittnerula* Hall & Clarke, and the evolution of cementation in the Brachiopoda. *Geological Magazine* 104:155–159.

Currey, J. 1970. *Animal Skeletons*. Studies in Biology no. 22. London: Arnold.

Dafni, J. 1986. A biomechanical model for the morphogenesis of regular echinoid tests. *Paleobiology* 12:143–160.

Davaud, E., and R. Wernli. 1974. Simulation de sections orientées de foraminifères Planispiralés au moyen d'un ordinateur. *Eclogae geologicae Helveticae* 67:31–38.

Davoli, F., and F. Russo. 1974. Una metodologia paleontometrica basata sul modello di Raup: Verifica sperimentale su rappresentanti follili del gen. *Subula* Schumacher. *Bollettino della Società paleontologica italiana* 13: 108–121.

Dawkins, R. 1987. *The Blind Watchmaker.* New York: Norton.

Dennett, D.C. 1996. *Darwin's Dangerous Idea.* New York: Simon & Schuster.

De Renzi, M. 1988. Shell coiling in some larger foraminifera: General comments and problems. *Paleobiology* 14:387–400.

De Renzi, M. 1995. Theoretical morphology of logistic coiling exemplified by tests of genus *Alveolina* (larger foraminifera). *Neues Jahrbuch für Geologie und Paläontologie, Abhandlungen* 195:241–251.

Dobzhansky, T. 1970. *Genetics of the Evolutionary Process.* New York: Columbia University Press.

Doescher, R. A. 1989. *Directory of Paleontologists of the World.* Lawrence: International Paleontological Association and University of Kansas Paleontological Institute.

Dommergues, J.-L., B. Laurin, and C. Meister. 1996. Evolution of ammonoid morphospace during the Early Jurassic radiation. *Paleobiology* 22:219–240.

Ebbesson, S. D. 1984. Evolution and ontogeny of neural circuits. *Behavioral and Brain Sciences* 7:321–366.

Ede, D. A., and J. T. Law. 1969. Computer simulation of vertebrate limb morphogenesis. *Nature* 221:244–248.

Ekaratne, S. U. K., and D. J. Crisp. 1983. A geometrical analysis of growth in gastropod shells, with particular reference to turbinate forms. *Journal of the Marine Biological Association of the United Kingdom* 63:777–797.

Ellers, O. 1993. A mechanical model of growth in regular sea urchins: Predictions of shape and a developmental morphospace. *Proceedings of the Royal Society of London* B254:123–129.

Erickson, R. O. 1982. Mathematical models of plant morphogenesis. *Acta Biotheoretica* 31A:132–151.

Ermentrout, B., J. Campbell, and G. Oster. 1986. A model for shell patterns based on neural activity. *Veliger* 28:369–388.

Fisher, D.C. 1985. Evolutionary morphology: Beyond the analogous, the anecdotal, and the ad hoc. *Paleobiology* 11:120–138.

Foote, M. 1991. Analysis of morphological data. In N. L. Gilinsky and P. W. Signor, eds., *Analytical Paleobiology.* Short Courses in Paleontology no. 4:59–86. Knoxville: University of Tennessee and the Paleontological Society.

Foote, M. 1992. Rarefaction analysis of morphological and taxonomic diversity. *Paleobiology* 18:1–16.

Foote, M. 1993. Discordance and concordance between morphological and taxonomic diversity. *Paleobiology* 19:185–204.

Foote, M. 1995. Morphological diversification of Paleozoic crinoids. *Paleobiology* 21:273–299.

Foote, M. 1997. The evolution of morphological diversity. *Annual Review of Ecology and Systematics* 28:129–152.

Foote, M., and R. H. Cowie. 1988. Developmental buffering as a mechanism for stasis: Evidence from the pulmonate *Theba pisana. Evolution* 42: 396–399.

Foote, M., and S. J. Gould. 1992. Cambrian and Recent morphological disparity. *Science* 258:1816.

Ford, E. D., A. Avery, and R. Ford. 1990. Simulation of branch growth in the *Pinaceae:* Interactions of morphology, phenology, foliage productivity, and the requirement for structural support, on the export of carbon. *Journal of Theoretical Biology* 146:15–36.

Fortey, R. A. 1983. Geometric constraints in the construction of grapholite stipes. *Paleobiology* 9:116–125.

Fortey, R. A. 1989. The collection connection. *Nature* 342:303.

Fortey, R. A., and A. Bell. 1987. Branching geometry and function of multiramous graptoloids. *Paleobiology* 13:1–19.

Fowler, D. R., H. Meinhardt, and P. Prusinkiewicz. 1992. Modeling seashells. *Computer Graphics* 26:379–387.

Fukutomi, T. 1953. A general equation indicating the regular forms of Mollusca shells, and its application to geology, especially in palaeontology. *Hokkaido University Geophysical Bulletin* 3:63–82.

Gardiner, A. R., and P. D. Taylor. 1980. Computer modelling of colony growth in a uniserial bryozoan. *Journal of the Geological Society of London* 137:107.

Gardiner, A. R., and P. D. Taylor. 1982. Computer modelling of branching growth in the bryozoan *Stromatopora. Neues Jahrbuch für Geologie und Paläontologie, Abhandlungen* 163:389–416.

Gardner, M. 1970. Mathematical games: The fantastic combinations of John Conway's new solitaire game "life." *Scientific American* 223 (October): 120–123.

Gardner, M. 1971. Mathematical games: On cellular automata, self-reproduction, the Garden of Eden and the game "life." *Scientific American* 224 (February):112–117.

Goodfriend, G. A. 1983. Some new methods for morphometric analysis of gastropod shells. *Malacological Review* 16:79–86.

Gould, S. J. 1970. Evolutionary paleontology and the science of form. *Earth Science Reviews* 6:77–119.

Gould, S. J. 1989. *Wonderful Life.* New York: Norton.

Gould, S. J. 1991. The disparity of the Burgess Shale arthropod fauna and the limits of cladistic analysis: Why we must strive to quantify morphospace. *Paleobiology* 17:411–423.

Gould, S. J. 1993. How to analyze Burgess Shale disparity—A reply to Ridley. *Paleobiology* 19:522–523.

Gould, S. J. 1995. A task for Paleobiology at the threshold of majority. *Paleobiology* 21:1–14.

Gould, S. J., and M. Katz. 1975. Distribution of ideal geometry in the growth of receptaculitids: A natural experiment in theoretical morphology. *Paleobiology* 1:1–20.

Gould, S. J., and R. C. Lewontin. 1979. The spandrels of San Marco and the panglossian paradigm: A critique of the adaptationist programme. *Proceedings of the Royal Society of London* B 205:581–598.

Graus, R. R. 1974. Latitudinal trends in the shell characteristics of marine gastropods. *Lethaia* 7:303–314.

Graus, R. R., and I. G. Macintyre. 1976. Light control of growth form in colonial reef corals: Computer simulation. *Science* 193:895–897.

Green, P. B., and R. S. Poethig. 1982. Biophysics of the extension and initiation of plant organs. In S. Subtelny and P. B. Green, eds., *Developmental Order: Its Origin and Regulation*, pp. 485–509. New York: Alan R. Liss.

Hall, B. K. 1996. *Baupläne*, phylotypic stages, and constraint: Why there are so few types of animals. In M. K. Hecht and others, eds., *Evolutionary Biology* 29:215–261.

Hallé, F., R. A. A. Oldeman, and P. B. Tomlinson. 1978. *Tropical Trees and Forests: An Architectural Analysis*. Berlin: Springer-Verlag.

Hallers-Tjabbes, C. C. 1979. Sexual dimorphism in *Buccinum undatum* L. *Malacologia* 18:13–17.

Harasewych, M. G. 1982. Mathematical modeling of the shells of higher prosobranchs. *Bulletin of the American Malacological Union, Inc., for* 1981:6–10.

Harrison, L. G. 1987. What is the status of reaction-diffusion theory thirty-four years after Turing? *Journal of Theoretical Biology* 125:369–384.

Hayami, I., and T. Okamoto. 1986. Geometric regularity of some oblique sculptures in pectinid and other bivalves: Recognition by computer simulations. *Paleobiology* 12:433–449.

Heath, D. J. 1985. Whorl overlap and the economical construction of the gastropod shell. *Biological Journal of the Linnean Society* 24:165–174.

Hickman, C. S. 1980. Gastropod radulae and the assessment of form in evolutionary paleontology. *Paleobiology* 6:276–296.

Hickman, C. S. 1988. Analysis of form and function in fossils. *American Zoologist* 28:775–793.

Hickman, C. S. 1993a. Biological diversity: Elements of a paleontological agenda. *Palaios* 8:309–310.

Hickman, C. S. 1993b. Theoretical design space: A new program for the analysis of structural diversity. In A. Seilacher and K. Chinzei, eds., *Progress in Constructional Morphology. Neues Jahrbuch für Geologie und Paläontologie, Abhandlungen* 190:169–182.

Hofmann, H. J. 1994. Quantitative stromatolitology. *Journal of Paleontology* 68:704–709.

Honda, H. 1971. Description of the form of trees by the parameters of the tree-like body: Effects of the branching angle and the branch length on the shape of the tree-like body. *Journal of Theoretical Biology* 31:331–338.

Honda, H., and J. B. Fisher. 1978. Tree branch angle: Maximizing effective leaf area. *Science* 199:888–890.

Honda, H., P. B. Tomlinson, and J. B. Fisher. 1981. Computer simulation of branch interaction and regulation by unequal flow rates in botanical trees. *American Journal of Botany* 68:569–585.

Horton, R. E. 1945. Erosional development of streams and their drainage basins: Hydrophysical approach to quantitative morphology. *Geological Society of America Bulletin* 56:275–370.

Hutchinson, J. M. C. 1989. Control of gastropod shell shape: The role of the preceding whorl. *Journal of Theoretical Biology* 140:431–444.

Huxley, J. S. 1932. *Problems of Relative Growth.* London: MacVeagh.

Illert, C. 1983. The mathematics of gnomonic seashells. *Mathematical Bioscience* 63:21–56.

Illert, C. 1987. Formulation and solution of the classical seashell problem. I. Seashell geometry. *Il Nuovo cimento* 9D:791–813.

Illert, C. 1989. Formulation and solution of the classical seashell problem. II. Tubular three-dimensional seashell surfaces. *Il Nuovo cimento* 11D: 761–780.

Illert, C. 1991. Foundations of theoretical conchology: From self-similarities in non-conservation mechanics. *Hadronic Journal Supplement* 6(4):361–473.

Inoué, S. 1982. The role of self-assembly in the generation of biological form. In S. Subtelny and P. B. Green, eds., *Developmental Order: Its Origin and Regulation*, pp. 35–76. New York: Alan R. Liss.

Jablonski, D. 1994. Extinctions in the fossil record. *Philosophical Transactions of the Royal Society of London, Series B*, 344:11–17.

Jacobson, G. G. 1980. Computer modelling of morphogenesis. *American Zoologist* 20:669–677.

Jaffe, L. F. 1982. Developmental currents, voltages, and gradients. In S. Subtelny and P. B. Green, eds., *Developmental Order: Its Origin and Regulation*, pp. 183–215. New York: Alan R. Liss.

James, M. A., A. D. Ansell, M. J. Collins, G. B. Curry, L. S. Peck, and M. C. Rhodes. 1972. Biology of living brachiopods. *Advances in Marine Biology* 28:175–387.

Jean, R. V. 1984. *Mathematical Approach to Pattern and Form in Plant Growth.* New York: Wiley.

Jean, R. V. 1994. *Phyllotaxis: A Systematic Study in Plant Morphogenesis.* Cambridge: Cambridge University Press.

Johnson, M. R., R. E. Tabachnick, and F. L. Bookstein. 1991. Landmark-based morphometrics of spiral accretionary growth. *Paleobiology* 17:19–36.

Kauffman, S. A. 1993. *The Origins of Order.* Oxford: Oxford University Press.

Kauffman, S. A. 1995. *At Home in the Universe.* Oxford: Oxford University Press.

Kendrick, D.C. 1993. Computer modelling of crinoid calyx morphologies and comparisons with real forms. *Geological Society of America, Abstracts with Programs* 25:A103.

Kendrick, D.C. 1996. Morphospace filling in flexible crinoids. In J. E. Repetski, ed., *Sixth North American Paleontological Convention Abstracts of Papers.* Paleontological Society Special Publication no. 8:208. Washington, D.C.: Smithsonian Institution and the Paleontological Society.

Kershaw, S., and R. Riding. 1978. Parameterization of stromatoporoid shape. *Lethaia* 11:233–242.

Kohn, A. J., and A. C. Riggs. 1975. Morphometry of the *Conus* shell. *Systematic Zoology* 24:346–359.

Labandeira, C., and N. C. Hughes. 1994. Biometry of the Late Cambrian trilobite genus *Dikelocephalus* and its implications for trilobite systematics. *Journal of Paleontology* 68:492–517.

Lauder, G. V. 1982. Introduction to *Form and Function: A Contribution to the History of Animal Morphology*, by E. S. Russell, pp. xi–xlv. Chicago: University of Chicago Press.

Leopold, L. B. 1971. Trees and streams: The efficiency of branching patterns. *Journal of Theoretical Biology* 31:339–354.

Levin, H. L. 1996. *The Earth Through Time.* Orlando, Fla.: Saunders College Publishing.

Lewontin, R. C. 1978. Adaptation. *Scientific American* 239 (September): 212–230.

Lindenmayer, A. 1968. Mathematical models for cellular interaction in development. Parts I and II. *Journal of Theoretical Biology* 18:280–315.

Linsley, R. M. 1977. Some "laws" of gastropod shell form. *Paleobiology* 3: 196–206.

Lison, L. 1942. Charactéristiques géométriques naturelles des coquilles des Lamellibranches. *Bulletin de l'Academie royale classe des sciences* 28: 377–390.

Lison, L. 1949. Recherches sur la form et la mechanique de dévelopment des coquilles des Lamellibranches. *Mémoires de l'Institut royale des sciences naturelles de Belgique.* 2er série 34:1–87.

Løvtrup, S., and M. Løvtrup. 1988. The morphogenesis of molluscan shells: A mathematical account using biological parameters. *Journal of Morphology* 197:53–62.

Løvtrup, S., and B. von Sydow. 1974. D'Arcy Thompson's theorems and the shape of the molluscan shell. *Bulletin of Mathematical Biology* 36:567–575.

Løvtrup, S., and B. von Sydow. 1976. An addendum to "D'Arcy Thompson's theorems and the shape of the molluscan shell." *Bulletin of Mathematical Biology* 38:321–322.

Lugar, L. 1990. Morphology. In D. E. G. Briggs and P. R. Crowther, eds., *Palaeobiology: A Synthesis*, pp. 307–313. Oxford: Blackwell Scientific Publications.

Lutz, T. M., and G. E. Boyajian. 1995. Fractal geometry of ammonoid sutures. *Paleobiology* 21:329–342.

McCartney, K., J. Ernisse, and D. E. Loper. 1994. Mathematical modeling of silicoflagellate skeletal morphology and implications concerning skeletal latticeworks. *Memoirs of the California Academy of Sciences* 17:87–94.

McCartney, K., and D. E. Loper. 1989. Optimized skeletal morphologies of silicoflagellate genera *Dictyocha* and *Distephanus*. *Paleobiology* 15:283–298.

McCartney, K., and D. E. Loper. 1992. Optimal models of skeletal morphology for the silicoflagellate genus *Corbisema*. *Micropaleontology* 38:87–93.

McGhee, G. R., Jr. 1978a. Analysis of the shell torsion phenomenon in the Bivalvia. *Lethaia* 11:315–329.

McGhee, G. R., Jr. 1978b. Geometric analysis of shell morphology in the biconvex articulate brachiopods. *Geological Society of America, Abstracts with Programs* 10(7):453.

McGhee, G. R., Jr. 1979a. Geometric analysis of biconvex brachiopod shell morphology: Ordinal distributions and stability strategies. *Geological Society of America, Abstracts with Programs* 11(1):44.

McGhee, G. R., Jr. 1979b. The geometry of biconvex brachiopod evolution. *Geological Society of America, Abstracts with Programs* 11(7):475.

McGhee, G. R., Jr. 1980a. Geometry of non-biconvex shell form in the Strophomenida and Orthida (Brachiopoda). *Geological Society of America, Abstracts with Programs* 12(7):479.

McGhee, G. R., Jr. 1980b. Shell form in the biconvex articulate Brachiopoda: A geometric analysis. *Paleobiology* 6:57–76.

McGhee, G. R., Jr. 1980c. Shell geometry and stability strategies in the biconvex Brachiopoda. *Neues Jahrbuch für Geologie und Paläontologie, Monatshefte* 1980(3):155–184.

McGhee, G. R., Jr. 1981. Mathematical models of bivalved shell growth and form. *Geological Society of America, Abstracts with Programs* 13(3):165.

McGhee, G. R., Jr. 1982a. The problematica *Palaeoxyris, Vetacapsula,* and *Fayolia:* A morphological comparison with recent chondrichthyan egg cases. *Proceedings of the Third North American Paleontological Convention* 2: 365–369.

McGhee, G. R., Jr. 1982b. Stability adaptations and shell form in the Brachiopoda. *Geological Society of America, Abstracts with Programs* 14(1,2): 39–40.

McGhee, G. R., Jr. 1989. Catastrophes in the history of life. In K. C. Allen and D. E. G. Briggs, eds., *Evolution and the Fossil Record,* pp. 26–50. London: Belhaven Press; Washington, D.C.: Smithsonian Institution Press.

McGhee, G. R., Jr. 1991. Theoretical morphology: The concept and its applications. In N. L. Gilinsky and P. W. Signor, eds., *Analytical Paleobiology.* Short Courses in Paleontology no. 4, pp. 87–102. Knoxville: University of Tennessee and the Paleontological Society.

McGhee, G. R., Jr. 1994. Comets, asteroids, and the Late Devonian mass extinction. *Palaios* 9:513–515.

McGhee, G. R., Jr. 1995a. Geometry of evolution in the biconvex Brachiopoda: Morphological effects of mass extinction. *Neues Jahrbuch für Geologie und Paläontologie, Abhandlungen* 197:357–382.

McGhee, G. R., Jr. 1995b. Morphological effects of mass extinction in the biconvex Brachiopoda: A geometric analysis. *Geological Society of America, Abstracts with Programs* 27(6):A370.

McGhee, G. R., Jr. 1996. *The Late Devonian Mass Extinction.* New York: Columbia University Press.

McGhee, G. R., Jr. 1999a. Morphological consequences of mass extinction in brachiopods: A theoretical perspective. In E. Savazzi, ed., *Functional Morphology of the Invertebrate Skeleton* (in press). New York: Wiley.

McGhee, G. R., Jr. 1999b. The optimum biconvex brachiopod in the theoretical spectrum of shell form. In E. Savazzi, ed., *Functional Morphology of the Invertebrate Skeleton* (in press). New York: Wiley.

McGhee, G. R., Jr. 1999c. Stability strategies and ordinal shell form variations in brachiopods: A theoretical perspective. In E. Savazzi, ed., *Functional Morphology of the Invertebrate Skeleton* (in press). New York: Wiley.

McGhee, G. R., Jr. 1999d. The theoretical enigma of non-biconvex brachiopod shell form. In E. Savazzi, ed., *Functional Morphology of the Invertebrate Skeleton* (in press). New York: Wiley.

McGhee, G. R., Jr., U. Bayer, and A. Seilacher. 1978. Shell torsion in bivalves. In A. Seilacher and R. Westphal, eds., *Report 1976–1978: Sonderforschungsbereich 53 Tübingen. Neues Jahrbuch für Geologie und Paläontologie, Abhandlungen* 157:64–65.

McGhee, G. R., Jr., U. Bayer, and A. Seilacher. 1991. Biological and evolutionary responses to transgressive-regressive cycles. In G. Einsele, W. Ricken, and A. Seilacher, eds., *Cycles and Events in Stratigraphy*, pp. 696–708. Berlin: Springer-Verlag.

McGhee, G. R., Jr., and E. S. Richardson Jr. 1982. First occurrence of the problematical fossil *Vetacapsula* in North America. *Journal of Paleontology* 56:1295–1296.

McKinney, F. K. 1981. Planar branch systems in colonial suspension feeders. *Paleobiology* 7:344–354.

McKinney, F. K., and D. M. Raup. 1982. A turn in the right direction: Simulation of erect spiral growth in the bryozoans *Archimedes* and *Bugula*. *Paleobiology* 8:101–112.

McKinney, M. L. 1984. Allometry and heterochrony in an Eocene echinoid lineage: Morphological change as a by-product of size selection. *Paleobiology* 10:407–419.

MacLeod, N. 1996. Empirical shape space representations and shape modeling of fossils from landmark-registered 2D outlines, 3D outlines, and 3D surfaces, with a comment on the indeterminacy of empirical "monomorphospace" analysis. In J. E. Repetski, ed., *Sixth North American Paleontological Convention Abstracts of Papers*. Paleontological Society Special Publication no. 8:254. Washington, D.C.: Smithsonian Institution and the Paleontological Society.

McMahon, T. A., and R. E. Kronauer. 1976. Tree structures: Deducing the principle of mechanical design. *Journal of Theoretical Biology* 59:443–466.

McNair, C. G., W. M. Kier, P. D. LaCroix, and R. M. Linsley. 1981. The functional significance of aperture form in gastropods. *Lethaia* 14:63–70.

McShea, D. W. 1993. Arguments, tests, and the Burgess Shale—A commentary on the debate. *Paleobiology* 19:399–402.

Mandelbrot, B. B. 1983. *The Fractal Geometry of Nature.* New York: Freeman.

Meinhardt, H. 1982. *Models of Biological Pattern Formation.* New York: Academic Press.

Meinhardt, H. 1984. A model for positional signalling, the threefold subdivision of segments and the pigmentation patterns of molluscs. *Journal of Embryology and Experimental Morphology* 83(supp.):101–112.

**301**

Meinhardt, H. 1995. *The Algorithmic Beauty of Sea Shells*. New York: Springer-Verlag.

Meinhardt, H., and M. Klinger. 1987. A model for pattern formation on the shells of molluscs. *Journal of Theoretical Biology* 126:63–89.

Moore, A. M., and O. Ellers. 1993. A functional morphospace, based on dimensionless numbers, for a circumferential, calcite, stabilizing structure in sand dollars. *Journal of Theoretical Biology* 162:253–266.

Morita, R. 1991a. Finite element analysis of a double membrane tube (DMS-tube) and its implication for gastropod shell morphology. *Journal of Morphology* 207:81–92.

Morita, R. 1991b. Mechanical constraints on aperture form in gastropods. *Journal of Morphology* 207:93–102.

Morita, R. 1993. Developmental mechanics of retractor muscles and the "dead spiral model" in gastropod shell morphogenesis. In A. Seilacher and K. Chinzei, eds., *Progress in Constructional Morphology. Neues Jahrbuch für Geologie und Paläontologie, Abhandlungen* 190:191–217.

Moseley, H. 1838. On the geometrical forms of turbinated and discoid shells. *Royal Society of London Philosophical Transactions for* 1838:351–370.

Moseley, H. 1842. On conchyliometry. *Philosophical Magazine* 21:300–305.

Naumann, C. F. 1840a. Beitrag zur Conchyliometrie. *Annalen der Physik* 50:223–236.

Naumann, C. F. 1840b. Über die Spiralen der Ammoniten. *Annalen der Physik* 51:245–259.

Naumann, C. F. 1845. Über die wahre Spirale der Ammoniten. *Annalen der Physik* 64:538–543.

Newkirk, G. F., and R. W. Doyle. 1975. Genetic analysis of shell-shape variation in *Littorina saxatilis* on an environmental cline. *Marine Biology* 30:227–237.

Nijhout, H. F. 1991. *The Development and Evolution of Butterfly Wing Patterns*. Washington, D.C.: Smithsonian Institution Press.

Niklas, K. J. 1982. Computer simulations of early land plant branching morphologies: Canalization of patterns during evolution? *Paleobiology* 8:196–210.

Niklas, K. J. 1986. Computer-simulated plant evolution. *Scientific American* 254 (March):78–86.

Niklas, K. J. 1994. Morphological evolution through complex domains of fitness. *Proceedings of the National Academy of Sciences (USA)* 91:6772–6779.

Niklas, K. J. 1997a. Effects of hypothetical developmental barriers and abrupt environmental changes on adaptive walks in a computer-generated domain for early vascular land plants. *Paleobiology* 23:63–76.

Niklas, K. J. 1997b. *The Evolutionary Biology of Plants*. Chicago: University of Chicago Press.

Niklas, K. J., and V. Kerchner. 1984. Mechanical and photosynthetic constraints on the evolution of plant shape. *Paleobiology* 10:79–101.

Odell, G. M., G. Oster, P. Alberch, and B. Burnside. 1981. The mechanical basis of morphogenesis I. Epithelial folding and invagination. *Developmental Biology* 85:446–462.

Okamoto, T. 1984. Theoretical morphology of *Nipponites* (a heteromorph ammonoid). *Kaseki* 36:37–51. (In Japanese: The journal is *Fossils*, published by the Palaeontological Society of Japan.)

Okamoto, T. 1988a. Analysis of heteromorph ammonoids by differential geometry. *Palaeontology* 31:35–52.

Okamoto, T. 1988b. Changes in life orientation during the ontogeny of some heteromorph ammonoids. *Palaeontology* 31:281–294.

Okamoto, T. 1988c. Developmental regulation and morphological saltation in the heteromorph ammonite *Nipponites*. *Paleobiology* 14:602–609.

Okamoto, T. 1993. Theoretical modelling of ammonite morphogenesis. In A. Seilacher and K. Chinzei, eds., *Progress in Constructional Morphology. Neues Jahrbuch für Geologie und Paläontologie, Abhandlungen* 190: 183–190.

Okamoto, T. 1996. Theoretical modeling of ammonoid morphology. In N. H. Landman, K. Tanabo, and R. A. Davis, eds., *Ammonoid Paleobiology: Topics in Geobiology No. 13*. New York: Plenum Press.

Oster, G. F., J. D. Murray, and A. K. Harris. 1983. Mechanical aspects of mesenchymal morphogenesis. *Journal of Embryology and Experimental Morphology* 78:83–125.

Oster, G. F., J. D. Murray, and P. Maini. 1985. A model for chondrogenic condensations in the developing limb: the role of extracellular matrix and cell tractions. *Journal of Embryology and Experimental Morphology* 89:93–112.

Papentin, F. 1973a. A Darwinian evolutionary system. I. Definition and basic properties. *Journal of Theoretical Biology* 39:397–415.

Papentin, F. 1973b. A Darwinian evolutionary system. III. Experiments on the evolution of feeding patterns. *Journal of Theoretical Biology* 39:431–445.

Papentin, F., and H. Röder. 1975. Feeding patterns: The evolution of a problem and a problem of evolution. *Neues Jahrbuch für Geologie und Paläontologie, Monatshefte* 1975(3):184–191.

Pickover, C. A. 1989. A short recipe for seashell synthesis. *IEEE Computer Graphics and Applications* 9 (November):8–11.

Pickover, C. A. 1991. *Computers and the Imagination*. New York: St. Martin's Press.

Prusinkiewicz, P., and D. R. Fowler. 1995. Chapter 10: Shell models in three dimensions. In H. Meinhardt, *The Algorithmic Beauty of Sea Shells*, pp. 162–181. Berlin: Springer-Verlag.

Prusinkiewicz, P., and A. Lindenmayer. 1990. *The Algorithmic Beauty of Plants*. Berlin: Springer-Verlag.

Prusinkiewicz, P., and A. Lindenmayer. 1996. *The Algorithmic Beauty of Plants* (reprint ed.). Berlin: Springer-Verlag.

Raup, D. M. 1961. The geometry of coiling in gastropods. *Proceedings of the National Academy of Sciences (USA)* 47:602–609.

Raup, D. M. 1962. Computer as aid in describing form in gastropod shells. *Science* 138:150–152.

## REFERENCES

Raup, D. M. 1966. Geometric analysis of shell coiling: General problems. *Journal of Paleontology* 40:1178–1190.

Raup, D. M. 1967. Geometric analysis of shell coiling: Coiling in ammonoids. *Journal of Paleontology* 41:43–65.

Raup, D. M. 1968. Theoretical morphology of echinoid growth. *Journal of Paleontology* 42:50–63.

Raup, D. M. 1969. Modeling and simulation of morphology by computer. *Proceedings of the North American Paleontological Convention, Part B:* 71–83.

Raup, D. M. 1972. Approaches to morphologic analysis. In T. J. M. Schopf, ed., *Models in Paleobiology,* pp. 28–44. San Francisco: Freeman, Cooper.

Raup, D. M., and J. A. Chamberlain Jr. 1967. Equations for volume and center of gravity in ammonoid shells. *Journal of Paleontology* 41:566–574.

Raup, D. M., and R. R. Graus. 1972. General equations for volume and surface area of a logarithmically coiled shell. *Mathematical Geology* 4:307–316.

Raup, D. M., and A. Michelson. 1965. Theoretical morphology of the coiled shell. *Science* 147:1294–1295.

Raup, D. M., and A. Seilacher. 1969. Fossil foraging behavior: Computer simulation. *Science* 166:994–995.

Reif, W.-E. 1980. A model of morphogenetic processes in the dermal skeleton of elasmobranchs. *Neues Jahrbuch für Geologie und Paläontologie, Abhandlungen* 159:339–359.

Reif, W.-E., and D. B. Weishampel. 1991. Theoretical morphology, an annotated bibliography 1960–1990. *Courier Forschungsinstitut Senckenberg* 142: 1–140.

Rex, M. A., and K. J. Boss. 1976. Open coiling in recent gastropods. *Malacologia* 15:289–297.

Rice, S. H. 1998. The bio-geometry of mollusc shells. *Paleobiology* 24:133–149.

Richter, R. 1924. Flachseebeobachtungen zur Paläontologie und Geologie. 9. Zur Deutung rezenter und fossiler Mäander-Figuren. *Senckenbergiana* 6: 141–157.

Ridley, M. 1993. Analysis of the Burgess Shale. *Paleobiology* 19:519–521.

Rudwick, M. J. S. 1964. The inference of function from structure in fossils. *British Journal for the Philosophy of Science* 15:27–40.

Rudwick, M. J. S. 1968. Some analytic methods in the study of ontogeny in fossils with accretionary skeletons. *Paleontological Society Memoir* 2:35–69.

Rudwick, M. J. S. 1970. *Living and Fossil Brachiopods.* London: Hutchinson University Library.

Russell, E. S. [1916] 1982. *Form and Function: A Contribution to the History of Animal Morphology.* Reprint, Chicago: University of Chicago Press.

Sattler, R., ed., 1978. Theoretical plant morphology. *Acta Biotheoretica* 27: *Supplement Folia Biotheoretica No. 7,* 142 pp.

Saunders, W. B., and A. R. H. Swan. 1984. Morphology and morphologic diversity of mid-Carboniferous (Namurian) ammonoids in time and space. *Paleobiology* 10:195–228.

Saunders, W. B., and D. M. Work. 1996. Shell morphology and suture complexity in Upper Carboniferous ammonoids. *Paleobiology* 22:189–218.

Savazzi, E. 1985. SHELLGEN: A BASIC program for the modeling of molluscan shell ontogeny and morphogenesis. *Computers and Geosciences* 11: 521–530.

Savazzi, E. 1987. Geometric and functional constraints on bivalve shell morphology. *Lethaia* 20:293–306.

Savazzi, E. 1990a. Biological aspects of theoretical shell morphology. *Lethaia* 23:195–212.

Savazzi, E. 1990b. C programs for displaying shaded three-dimensional objects on a PC. *Computers and Geosciences* 16:195–209.

Savazzi, E. 1990c. Theoretical morphology of shells aided by microcomputers. In J. T. Hanley and D. F. Merriam, eds., *Microcomputer Applications in Geology II*, pp. 229–240. Oxford: Pergamon Press.

Savazzi, E. 1993. C++ classes for theoretical shell morphology. *Computers and Geosciences* 19:931–964.

Savazzi, E. 1995. Theoretical shell morphology as a tool in constructional morphology. *Neues Jahrbuch für Geologie und Paläontologie, Abhandlungen* 195:229–240.

Schindel, D. E. 1990. Unoccupied morphospace and the coiled geometry of gastropods: Architectural constraint or geometric covariation? In R. A. Ross and W. D. Allmon, eds., *Causes of Evolution*, pp. 270–304. Chicago: University of Chicago Press.

Schips, M. 1922. *Mathematik und Biologie*. Leipzig-Berlin: Teubner.

Scott, G. H. 1974. Biometry of the foraminiferal shell. In R. H. Hedley and C. G. Adams, eds., *Foraminifera*, vol. 1, pp. 55–151. London: Academic Press.

Seilacher, A. 1970. Arbeitskonzept zur Konstruktions-Morphologie. *Lethaia* 3:393–396.

Seilacher, A. 1973. Fabricational noise in adaptive morphology. *Systematic Zoology* 22:451–465.

Seilacher, A. 1991. Self-organizing mechanisms in morphogenesis and evolution. In N. Schmidt-Kittler and K. Vogel, eds., *Constructional Morphology and Evolution*, pp. 251–271. Berlin: Springer-Verlag.

Signes, M., J. Bijma, C. Hemleben, and R. Ott. 1993. A model for planktic foraminiferal shell growth. *Paleobiology* 19:71–91.

Signor, P. W. 1982. A critical re-evaluation of the paradigm method of functional inference. *Neues Jahrbuch für Geologie und Paläontologie, Abhandlungen* 164:59–63.

Silk, W. K., and M. Hubbard. 1991. Axial forces and normal distributed loads in twining stems of morning glory. *Journal of Biomechanics* 24:599–606.

Sneath, P. H. A. 1967. Trend-surface analysis of transformation grids. *Proceedings of the Zoological Society of London* 151:65–122.

Sneath, P. H. A., and R. R. Sokal. 1973. *Numerical Taxonomy*. San Francisco: Freeman.

Spivey, H. R. 1988. Shell morphometry in barnacles: Quantification of shape and shape change in *Balanus*. *Journal of Zoology* 216:265–294.

REFERENCES

Starcher, R. W. 1987. A constructional morphologic analysis of the fenestrate colony meshwork. Ph.D. diss., Rutgers University.

Stasek, C. R. 1963. Geometrical form and gnomonic growth in the bivalved Mollusca. *Journal of Morphology* 112:215–231.

Steinberg, M. S. 1970. Does differential adhesion govern self-assembly processes in histogenesis? Equilibrium configurations and the emergence of a hierarchy among populations of embryonic cells. *Journal of Experimental Zoology* 173:395–434.

Stevens, P. S. 1974. *Patterns in Nature.* Boston: Little, Brown.

Stone, J. R. 1995. CerioShell: A computer program designed to simulate variation in shell form. *Paleobiology* 21:509–519.

Stone, J. R. 1996a. Computer simulated shell shape and size variation in the Caribbean land snail genus *Cerion:* A test of geometrical constraints. *Evolution* 50:341–347.

Stone, J. R. 1996b. The evolution of ideas: A phylogeny of shell models. *American Naturalist* 148:904–929.

Stone, J. R. 1997a. Mathematical determination of coiled shell volumes and surface areas. *Lethaia* 30:213–219.

Stone, J. R. 1997b. The spirit of D'Arcy Thompson dwells in empirical morphospace. *Mathematical Biosciences* 142:13–30.

Strahler, A. N. 1952. Hypsometric (area-altitude) analysis of erosional topography. *Geological Society of America Bulletin* 63:1117–1142.

Strathmann, R. R. 1975. Limitations on diversity of forms: Branching of ambulacral systems of echinoderms. *American Naturalist* 109:177–190.

Strathmann, R. R. 1978. Progressive vacating of adaptive types during the Phanerozoic. *Evolution* 32:907–914.

Swan, A. R. H. 1990a. A computer simulation of evolution by natural selection. *Journal of the Geological Society, London* 147:223–228.

Swan, A. R. H. 1990b. Computer simulation of invertebrate morphology. In D. L. Bruton and D. A. T. Harper, eds., *Microcomputers in Palaeontology.* Contributions of the University of Oslo Palaeontological Museum no. 370:32–45. Oslo: University of Oslo.

Swan, A. R. H., and S. Kershaw. 1994. A computer model for skeletal growth of stromatoporoids. *Palaeontology* 37:409–423.

Swan, A. R. H., and W. B. Saunders. 1987. Function and shape in late Paleozoic (mid-Carboniferous) ammonoids. *Paleobiology* 13:297–311.

Telford, M. 1990. Computer simulation of deposit-feeding by sand dollars and sea biscuits (Echinoidea: Clypeasteroida). *Journal of Experimental Marine Biology and Ecology* 142:75–90.

Thom, R. 1975. *Structural Stability and Morphogenesis.* Reading, Mass.: W. A. Benjamin.

Thomas, R. D. K. 1979. Morphology, constructional. In R. W. Fairbridge and D. Jablonski, eds., *Encyclopedia of Paleontology; Encyclopedia of Earth Sciences,* vol. 7, pp. 482–487. Stroudsburg, Pa.: Dowden, Hutchinson & Ross.

Thomas, R. D. K., and W.-E. Reif. 1991. Design elements employed in the construction of animal skeletons. In N. Schmidt-Kittler and K. Vogel, eds., *Constructional Morphology and Evolution*, pp. 283–294. Berlin: Springer-Verlag.

Thomas, R. D. K., and W.-E. Reif. 1993. The skeleton space: A finite set of organic designs. *Evolution* 47:341–360.

Thompson, D'A. W. 1917. *On Growth and Form*. Cambridge: Cambridge University Press.

Thompson, D'A. W. 1942. *On Growth and Form*. 2d ed. Cambridge: Cambridge University Press.

Turing, A. M. 1952. The chemical basis of morphogenesis. *Philosophical Transactions of the Royal Society of London*, B237:37–72.

Urbanek, A., and J. Uchmański. 1990. Morphogenesis of uniaxiate graptoloid colonies—A mathematical model. *Paleobiology* 16:49–61.

Valentine, J. W. 1986. Fossil record of the origins of Baupläne and its implications. In D. M. Raup and D. Jablonski, eds., *Patterns and Processes in the History of Life*, pp. 209–222. Berlin: Springer-Verlag.

Van Valen, L. 1978. Arborescent animals and other colonoids. *Nature* 276:318.

Van Valkenburgh, B. 1985. Locomotor diversity within past and present guilds of large predatory mammals. *Paleobiology* 11:406–428.

Van Valkenburgh, B. 1988. Trophic diversity in past and present guilds of large predatory mammals. *Paleobiology* 14:155–173.

Verduin, A. 1982. How complete are diagnoses of coiled shells of regular build? A mathematical approach. *Basteria* 45:127–142.

Vermeij, G. J. 1971. Gastropod evolution and morphological diversity in relation to shell geometry. *Journal of Zoology* 163:15–23.

Vermeij, G. J. 1973a. Molluscs in mangrove swamps: Physiognomy, diversity, and regional differences. *Systematic Zoology* 22:609–624.

Vermeij, G. J. 1973b. West Indian molluscan communities in the rocky intertidal zone: A morphological approach. *Bulletin of Marine Science* 23:351–386.

Vermeij, G. J. 1975. Evolution and distribution of left-handed and planispiral coiling in snails. *Nature* 254:419–420.

Vogel, H. 1979. A better way to construct the sunflower head. *Mathematical Biosciences* 44:179–189.

Ward, P. 1980. Comparative shell shape distributions in Jurassic-Cretaceous ammonites and Jurassic-Tertiary nautilids. *Paleobiology* 6:32–43.

Waters, J. A. 1977. Quantification of shape by use of Fourier analysis: The Mississippian blastoid genus *Pentremites*. *Paleobiology* 3:288–299.

Watters, G. T. 1991. Utilization of a simple morphospace by polyplacophorans and its evolutionary implications. *Malacologia* 33:221–240.

Webb, L. P., and A. R. H. Swan. 1996. Estimation of parameters of foraminiferal test geometry by image analysis. *Palaeontology* 39:471–475.

Weishampel, D. B. 1991. A theoretical morphological approach to tooth replacement in lower vertebrates. In N. Schmidt-Kittler and K. Vogel, eds., *Constructional Morphology and Evolution*, pp. 295–310. Berlin: Springer-Verlag.

REFERENCES

Weliky, M., and G. Oster. 1990. The mechanical basis of cell rearrangement. I. Epithelial morphogenesis during *Fundulus* epiboly. *Development* 109: 373–386.

Whittaker, R. H. 1969. New concepts of kingdoms of organisms. *Science* 163: 150–160.

Williamson, P. G. 1981. Palaeontological documentation of speciation in Cenozoic molluscs from Turkana Basin. *Nature* 293:437–443.

Wills, M. A. 1996. A comparison of crustacean disparity and phylogeny through time. In J. E. Repetski, ed., *Sixth North American Paleontological Convention Abstracts of Papers*. Paleontological Society Special Publication no. 8:423. Washington, D.C.: Smithsonian Institution and the Paleontological Society.

Wills, M. A., D. E. G. Briggs, and R. A. Fortey. 1994. Disparity as an evolutionary index: A comparison between Cambrian and Recent arthropods. *Paleobiology* 20:93–130.

Willshaw, D. J., and C. von der Malsburg. 1976. How patterned neural connections can be set up by self-organization. *Proceedings of the Royal Society of London* B194:431–445.

Wilson, E. O., and W. H. Bossert. 1971. *A Primer of Population of Biology*. Sunderland: Sinauer.

Wolfram, S. 1984a. Cellular automata as models of complexity. *Nature* 311: 419–424.

Wolfram, S. 1984b. Preface, Los Alamos conference on cellular automata. *Physica (Series D)* 10:vii–xii.

Wolfram, S. 1984c. Universality and complexity in cellular automata. *Physica (Series D)* 10:1–35.

Woodger, J. H. 1945. On biological transformations. In W. E. Le Gros Clark and P. B. Medawar, eds., *Essays on Growth and Form Presented to D'Arcy Wentworth Thompson*, pp. 95–120. Cambridge: Cambridge University Press.

Wright, S. 1932. The roles of mutation, inbreeding, crossbreeding and selection in evolution. *Proceedings of the XI International Congress of Genetics* 1: 356–366.

# INDEX

Abelson, H., 256
Accretionary growth systems, evolution of morphogenetic models of, 245, 247–251; helicospiral bivalved theoretical morphospaces of, 112–130, 139–140; helicospiral univalved theoretical morphospaces of, 66–84, 92–111; mathematical simulation of, 209–225; morphogenetic models of, 226–251; planispiral bivalved theoretical morphospaces of, 112–139; planispiral univalved theoretical morphospaces of, 66–92
Ackerly, S.C., 26, 79, 90–91, 95, 110–111, 137–139, 221, 225, 227, 231, 236, 239–240, 248–250, 279; bivalved theoretical morphospace of, 123–126; univalved theoretical morphospace of, 82–84
Adaptive landscape, concept of, 10–14, 57, 73, 88–89, 182–189; Fujiyama adaptive landscape, 183–184; *NK* adaptive landscape, 183; random adaptive landscape, 183–

184; rugged adaptive landscape, 183
Aguado, R., 244–246, 248–250
Allometric functions, 233–234, 288
Anisometric growth, mathematical simulation of, 217–223, 225, 231–232, 242
Arrowheads, 19
Arthropods, 142, 181, 207–208, 286, 287
Avery, A., 255

Bard, J., 207
Bard, J. B. L., 264, 265, 272
Barnacles, 27, 29
Batschelet, E., 263, 266
Bauplan, concept of, 6–8, 288
Bayer, U., x, 82, 142, 147, 201, 204–205, 213, 223, 227, 230, 234, 237, 244, 248–250
Behavior, 272–276
Bell, A. D., 38–40, 253–255, 257
Berger, W. H., 148, 150, 151, 257, 259, 284; chambered theoretical morphospace of, 142–147

**309**